Catalogue of Central Peaks and Floor Objects of the Lunar Craters on the Visible Hemisphere

Catalogue of Central Peaks and Floor Objects of the Lunar Craters on the Visible Hemisphere

Compiled by

Naosuke SEKIGUCHI

Tokyo Astronomical Observatory, University of Tokyo

UNIVERSITY OF TOKYO PRESS
UNIVERSITY PARK PRESS

Printed in Japan.

Published jointly by
UNIVERSITY OF TOKYO PRESS
UNIVERSITY PARK PRESS
Baltimore • London • Tokyo

Library of Congress Cataloging in Publication Data

Sekiguchi, Naosuke, 1925–
Catalogue of central peaks and floor objects of the lunar craters on the visible hemisphere.

1. Lunar craters. I. Title.
QB591.S44 1972 919.91′02 70-38632
ISBN 0-8391-0649-1

Table of Contents

PREFACE

The development of the features of the moon's surface have attracted the interest of selenographers throughout history. The origin of ringed formations, ordinarily called craters, found in great numbers on the moon's surface have been a special point of dispute. Though these formations are common on the moon, they are rarely found on earth, and it is therefore very difficult to apply theories of development extrapolated from terrestrial topography to the lunar craters.

A partial solution to this problem is to investigate parts of the lunar terrain which are likely to be closely related to the development of these craters. Various objects found on the floors of lunar craters are suitable for this purpose, since it is obvious that the floor objects were formed in connection with the development of the craters.

In planning these investigations, I made a survey of lunar floor objects, especially the central peaks of craters, paying attention to the differences in type and distribution on the lunar globe. The present catalogue is the result of this survey.

This listing attempts to register all the floor objects which are observable by telescopic methods from the earth with greatest attention given to objects rising above the surface of the crater floors. Since my interests were originally the central peaks, there are some imperfections in the description and listing of depressed objects, such as pits, cracks and depressions.

In compiling the present catalogue, *Kuiper's Photographic Lunar Atlas* was my chief reference. In addition, I used many lunar photographs, published and unpublished, especially my own photographic and visual observations obtained using the 91 cm reflector at the Dodaira Station of the Tokyo Astronomical Observatory.

A description of the way compilation was carried out follows. First, I chose craters which contained any floor objects to be registered and measured their diameters and heights using *Kuiper's Photographic Lunar Atlas*. In this atlas, most craters are visible in several different photographs. Therefore, the sizes and heights were measured separately from individual photographs. The heights were calculated by measuring the length of the shadow cast by the objects. These measurements often did not agree with each other. I examined these values considering the sharpness of each photograph, the state of illumination and the terrain surrounding the craters and adopted what seemed to be the most reasonable values of the measured quantities. When the form of the objects deviated from the circular, their major and minor diameters were determined. In cases where the values of the measured heights from east and west bottoms were not in accord, both values were adopted. When it was difficult to get photographs enabling measurement of height or when lunar terrain around an object was not suitable to height measurement (for example, the floor was so deep that the object was submerged in the shadow of the wall when the sun was low enough to generate a shadow of the object), I used the term "Unmeasurably Low" (U.L.) or determined the upper limit of its height.

The photographs from space probes used in the course of this compilation were from the Ranger probes VII, VIII and IX. They supplied exactness of description to the present catalogue and added two dimple craters containing central objects to the catalogue.

I would like to thank Professor Z. Kopal of the University of Manchester for enabling me to use the facilities there to examine photographs of the Lunar Orbiter IV during my stay in Manchester. These photographs gave me the opportunity to examine the exactness of the descriptions of objects on the marginal zone, which are very difficult to observe from the earth. As a result of using these facilities, I was able to check details of these marginal objects and add some new objects.

Once the photographs from the lunar space probes are available, it will be easy to make a similar catalogue of the objects on the averted side of the moon, but since I wished to conclude ten years' work with this catalogue, I confined it to the visible lunar hemisphere. The compilation of a similar catalogue about the averted hemisphere can be expected in the near future.

In the course of the compilation of the present catalogue, Mrs. F. Miyamoto of the Tokyo Astronomical Observatory helped me to choose the objects from lunar photographs and to measure and calculate their sizes and heights. The sketches of lunar craters in the present catalogue were drawn by Mrs. F. Miyamoto and Miss K. Yamaguchi of the Dodaira Station, whom I would also like to thank.

I would like to express my gratitude to Mr. Patrick Moore for his valuable suggestions and warm encouragement of my compilation of this catalogue.

Explanatory Note

In the main part of this catalogue, the moon's visible hemisphere is divided into 25 sections according to the values of $\xi = \sin \lambda \cos \beta$ and $\eta = \sin \beta$, where λ and β are selenographic longitude and latitude, respectively. This manner of division is the same as that used in the book by Wilkins and Moore. In each section the craters were arranged in alphabetical order. Unnamed formations are designated by their positions relative to nearby named craters. According to the nomenclature adopted at the XIVth General Assembly of the I.A.U., some designations were changed. For example, "Hedin" was changed to "Unnamed Formation to the South of Olbers" and "Pingré S" to "Unnamed Formation to the Northwest of Bailly."

The parentheses after the designations of craters contain figures with three decimal places of ξ and η and, separated from them by a semicolon, the diameter of the crater in km. In cases where the difference between major and minor diameters is remarkable, both values are given, separated by an oblique line. These numbers serve to identify each formation. The values in parentheses are taken from the Arizona Catalogue, except in cases where the formation is markedly different from the one registered in the Arizona Catalogue. The coordinates and diameters of the dimple craters found by the space probes were determined independently.

Each object was listed according to the numbers in parentheses on the basis of its distinctiveness and sometimes its situation. In cases where the object has a Greek letter designation, the letter is written after the object number and separated from it by a colon. The diameter and height of the object in km are given in brackets and the two figures separated by a semicolon. When the form of the object deviated from the circular, its major and minor diameters are given with an oblique line between them. In the case of long objects, such as dykes, ridges or cracks, the length and breadth are given. When it is probable that the heights measured from east and west bottoms are different, "e" or "w" follow the height figure. When only the upper limit of height was measurable, an inequality sign (<) is written before the figure for the upper limit. When the height was too low to be measured, "U.L." (Unmeasurably Low) is given instead of a measurement.

In ordinary cases, the values in brackets are rounded after the final decimal place, but when the last figure is 0 or 5, I adopted a special convention. When the final figure is 0, the following decimal figure is larger than -2.5 and smaller than 2.5; when the final figure is 5, the following decimal figure is larger than 2.5 and smaller than 7.5. When the last figure 0 or 5 is used in the ordinary sense, I placed a dot above it. According to these conventions,

	$a=1$	means	$0.5<a<1.5$,
	$a=1.0$	means	$0.75<a<1.25$,
	$a=1.\dot{0}$	means	$0.95<a<1.05$,
	$a=1.5$	means	$1.25<a<1.75$,
and	$a=1.\dot{5}$	means	$1.45<a<1.55$.

These conventions are valid only for the figures in brackets.

In this catalogue, north refers to the direction in which the selenographic latitude β increases, the selenographic longitude λ remaining constant. However, in the picutres in this catalogue, the arrows always point in the direction in which the projected rectangular coordinate $\eta=\sin\beta$ increases, the $\xi=\sin\lambda\cos\beta$ remaining constant. These two directions are much different near the polar region. East is the direction toward Mare Crisium from the central meridian and west is toward Grimaldi. The arrows always represent 20 km, except in cases where special remarks are written.

The pictures in the present catalogue serve to identify each object in a crater when it contains more than one object. The contours of the peripheral wall represent the position of its ridge, the highest line of the wall. The hatch marks on the wall represent the shadows which show the direction of illumination. Objects on the floors are distinguished as protruding or depressed by the placement of the shadow. For example, in the figure below, the illumination comes from the left, therefore, the object on the left protrudes and the object on the right is depressed. This figure shows the position of objects relative to the peripheral wall, but it does not represent their forms correctly.

The types of registered objects are classified in the following manner:

(1) Mound or dome: rising terrain with a spherical boss shape.

(2) Elevation: faint rising terrain without a definite periphery.

(3) Peak: rising object with a tapered summit.

(4) Mountain or mountain block: lofty terrain with an irregular form.

(5) Hill: rising terrain whose peak is not steep or rough.

(6) Dyke or bank: long, narrow rising terrain.

(7) Scarp, cliff or crack: long, narrow depressed terrain.

(8) Ridge: linear object with a comparatively sharp top.

(9) Lineament: faint linear object.

(10) Platform or plateau: rising terrain with a broad and flat surface.

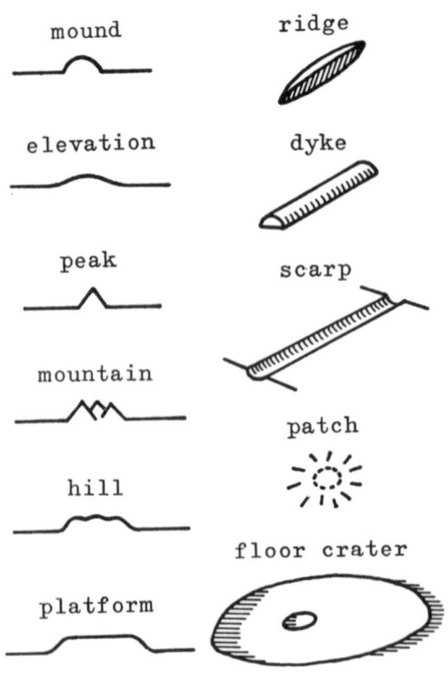

(11) Floor crater: small crater on the floor of a larger crater.
(12) Pit: circular hole whose periphery has no height compared with the surrounding plane.
(13) Patch: a small white or black area with the same elevation as the surrounding terrain.

In the present catalogue, I have paid special attention to some objects. One of these is the "true central peak." A typical example is peak (3) in Copernicus. It is not the most remarkable object among the group of hills on the crater floor, but it is situated in the center of the crater and has a regular form. The true central peak was probably formed during the final stage of activity of the crater. Furthermore, I paid special attention to the objects which seemed to have some relationship with the surrounding structure of the craters. There are many cases in which the objects on the floor seem to be related to a lattice or parallel structure around the crater. Terrain which penetrates the walls of craters is often in evidence; I have termed this "penetrating structures."

In the index of this catalogue, the first column shows the number of the section and crater and the second column gives the designation of the crater arranged in alphabetical order. The third to the fifth columns give the coordinates and diameters of the craters. These figures are the same as those in parentheses in the main body of the catalogue. The sixth column shows the kinds and numbers of registered objects in the crater and sometimes gives special remarks.

List of Abbreviations

Arizona Catalogue	: The system of Lunar Craters, Quadrant I–IV. D. W. G. Arthur et al. *Communications of the Lunar and Planetary Laboratory*, **2**, 71, 1963; **3**, 1, 1964; **3**, 61, 1965; **5**, 1, 1966.
c	: floor crater, pit.
ccc	: concentric crater.
cef	: centrally elevated floor.
cpc	: compound crater.
d	: dyke.
E.	: east.
e	: elevation.
h	: hill.
M	: mountain, mountain block.
m	: mound, dome.
mow	: mound on the inner slope of wall.
N.	: north.
P	: platform, plateau.
p	: peak.
ps	: penetrating structure.
pt	: patch.
r	: ridge.
rc	: ray crater.
rf	: rugged floor.
S.	: south.
s	: scarp, cliff, crack.
tcp	: true central peak.
U.F.	: unnamed formation.
U.L.	: unmeasurably low.
W.	: west.
Wilkins and Moore	: *The Moon*. H. P. Wilkins and Patrick Moore. London: Faber and Faber, 1954.

Catalogue of Central Peaks and Floor Objects of the Lunar Craters on the Visible Hemisphere

I

1. Agrippa (+182,+072;46)
   (1) ⌜10;1.4w/0.7e⌟ A large central peak, having non-conical form.
2. Albategnius (+070,-195;136)
   (1):α ⌜14;3.0⌟ The main body. An oblong ridge. There is a summit pit of elliptical form whose major and minor diameters are about 1.8 km and 1.0 km respectively, according to the Ranger IX photographs.
   (2) ⌜3;0.5⌟,(3) ⌜3;0.7⌟ Tiny hills on the north-east side of the (1). These peaks look as a part of crater wall remnant.

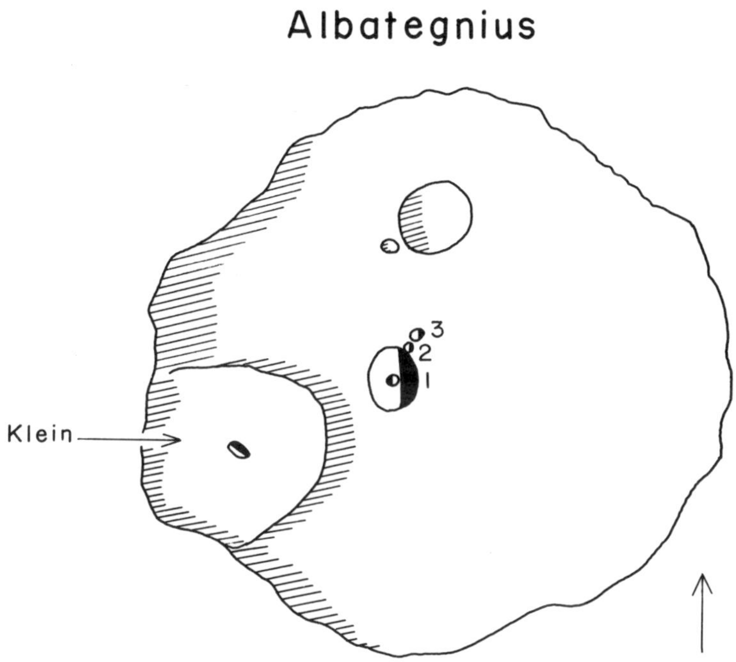

3. Anděl A (+192,-187;14)
   (1) ⌜2.5;0.4⌟ A central peak.
   (2) ⌜2;0.3⌟ Smaller hill on the north-west bottom of the (1).

4. Bode (-042,+117;19)
   (1) ⌜6;<0.8⌟ Very low mound. The floor is deeper towards the south.
5. Boscovich P (+175,+199;67/17)
   (1) ⌜2.7;0.2⌟ A mound on the floor of very curious form. Its coordinates are ξ = +0.190, amd η =+0.193. Its form is slightly oblong along the direction of linearments around this formation.
6. Flammarion W (-041,-037;7)
   (1) ⌜2;<0.1⌟ The central mountain has a ridge-like form, and seems as a part of fault structure through this crater. Its highest part is northerly eccentric.
7. Godin (+177,+032;35)
   (1) ⌜9;1.5w/0.9e⌟ A large central peak having a summit crater. The floor is very rugged.
8. Gyldén (+005,-093;47)
   (1) ⌜15;<1⌟ In the south-east part of the floor, there is a very low heap of rugged form, with a central crater, of which diameter is about 4 km.

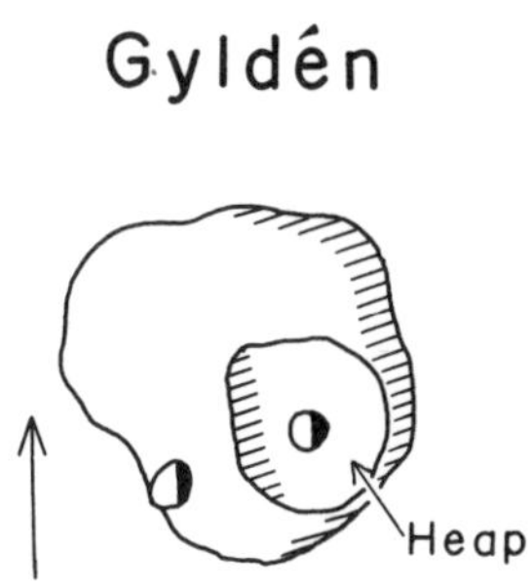

9. Halley (+100,-140;36)
   (1) ⌜3;<0.4⌟ Very low and small central peak on flat floor.

10. Herschel (-036,-099;41)

(1) ⌜8;0.8⌟ The central mountain has a form like assembly of three blocks, of which the south-west one is the highest. The floor is very deep (its depth is about 8 km) and has a flat part in the east.

Herschel

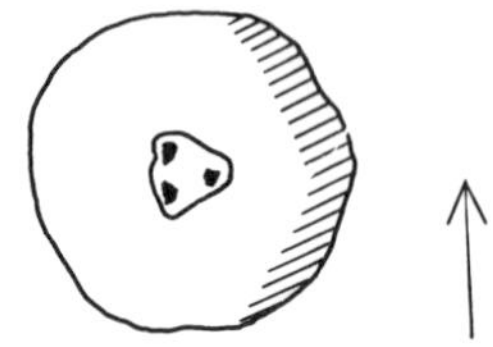

11. Hind (+127,-138;29)

(1) ⌜4;0.5w⌟ (2)⌜4;0.4w⌟ At north-westerly eccentric position, lie two blocks (1) and (2), which have steeper slopes towards west.

Hind

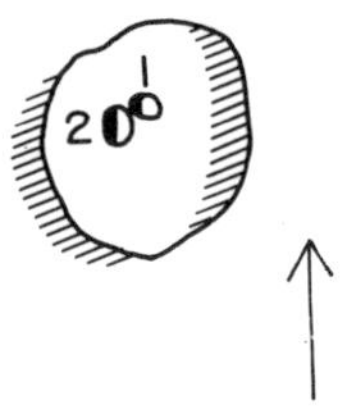

Hipparchus

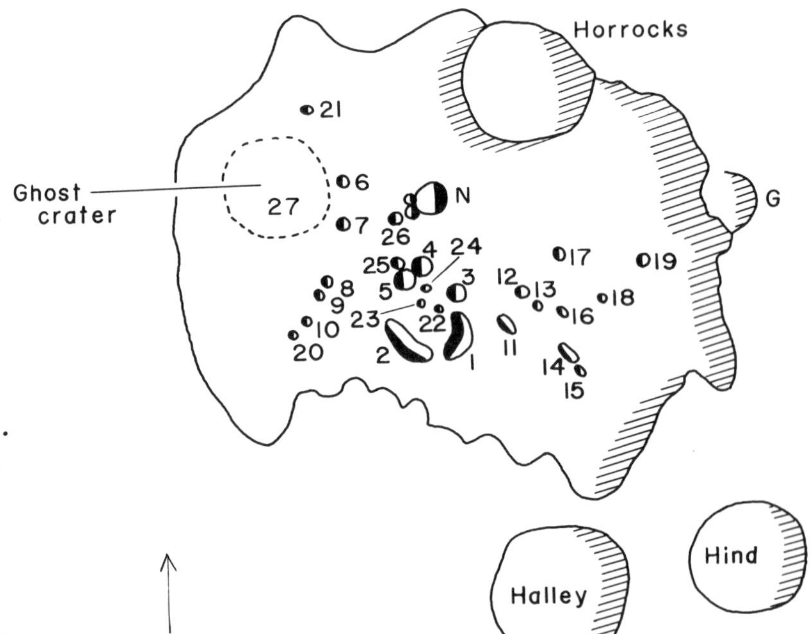

12. Hipparchus (+084, -096;151)

(1) ⌜11/3;1⌟ (2) ⌜8/3;1⌟ Ridges, which are remnants of an old crater, Hipparchus X.

(3) ⌜3;0.4⌟ A round mound.

(4) ⌜2.3;0.35⌟ (5) ⌜2;0.3⌟ Distinct peaks.

(6) ⌜1.5;0.1⌟ (7) ⌜2.3;0.1⌟ (8) ⌜2.0;0.1⌟ (9) ⌜1.6;0.1⌟ (10) ⌜<1;<0.1⌟ Mounds lying in a line.

(11) ⌜4/1;<0.1⌟ A faint ridge.

(12) ⌜2.5;0.3⌟ A distinct peak.

(13) ⌜1.7;0.1⌟ A tiny mound.

(14) ⌜4/1.7;0.4⌟ (15) ⌜2/0.2⌟ Ridges.

(16) ⌜<1;<0.1⌟ A very faint ridge.

(17) ⌜1;0.1⌟ (18) ⌜1;0.1⌟ (19) ⌜1;0.1⌟ Very faint mounds. On the south-east part of the floor, we can see a structure parallel to the grid system around Hipparchus.

(20) ⌜<1;<0.1⌟ (21) ⌜<0.8;<0.1⌟ Very faint mounds.

(22) ⌜<1;<0.1⌟ (23) ⌜<1;<0.1⌟ (24) ⌜<1;<0.1⌟ Low mounds.

(25) ⌜<0.8;<0.1⌟ (26) ⌜<0.8;<0.1⌟ Faint elevations.

(27) ⌜23;U. L.⌟ A ghost crater on the north-west part of the floor.

13. Hipparchus, U. F. on the W. Rampart of (+048,-094;6)

(1) ⌞2;<0.1⌟ A very low and tiny elevation at westerly eccentric position.

14. Hipparchus, U. F. on the N.-W. Rampart of (+057,-065;16)

(1) ⌞2; 0.1⌟ A very low round mound.

15. Hipparchus, U. F. to the N. of (+084,-045;8)

(1) ⌞2;<0.1⌟ A faint elevation at north-westerly eccentric position.

16. Hipparchus, U. F. on the S. Part of (+097,-120;10)

(1) ⌞2;0.1⌟ (2) ⌞2;0.1⌟ Two mounds stand along the linear structure of this region. The mound (2) is higher.

U.F. on the South Part of Hipparchus

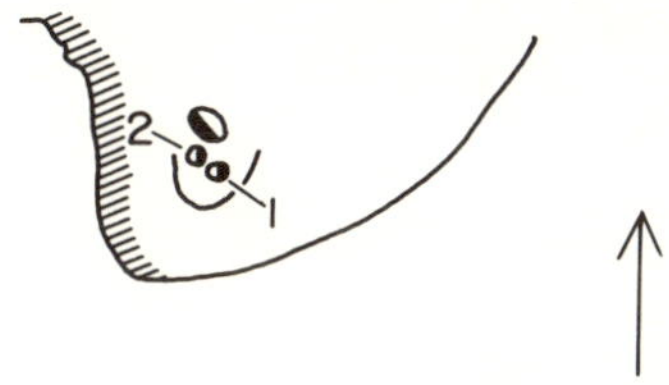

17. Horrocks (+102,-069;31)

(1) ⌞4;0.6w⌟ The floor is very rugged and seems like a crowd of low mounds. On south-westerly eccentric position, eminent mound is seen.

18. Hyginus S (+138,+112;29)

(1) ⌞<1;<0.1⌟ (2) ⌞<1;<0.1⌟ Two tiny domes stand in 4 km apart from each other.

Hyginus S

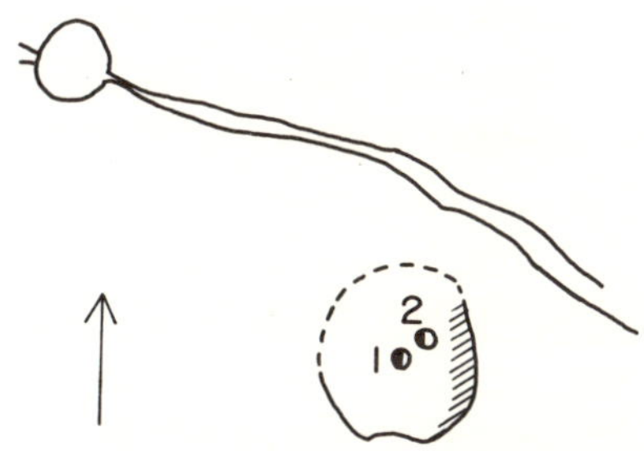

19. Klein, U. F. on the W. of (+027, -199; 34)

(1) ⌞10/5;0.7⌟ (2)⌞7/4;0.5̇⌟ (3)⌞3/2;0.3⌟ (4)⌞5̇/4;0.3⌟ A chain of mountains.

U.F. on the West of Klein

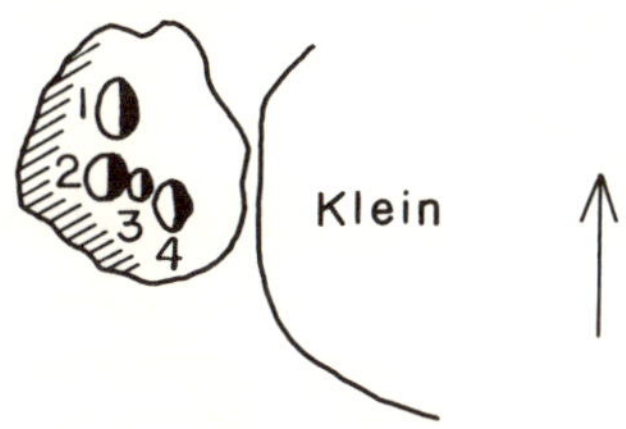

20. Lalande (-149,-078;24)

A ray crater.

(1) ⌞4;<0.2⌟ Very low central mound.

(2) ⌞3;0.4⌟ An eminent mound on the north wall.

Lalande

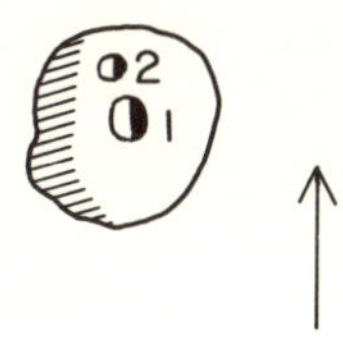

21. Mösting (-101,-012;26)

(1) ⌞6;<0.2⌟ This low and broad mound is slightly elevated part of floor, situated on southerly eccentric position.

(2) ⌞2;0.2⌟ A tiny hill on the north wall. This aspect is very similar as that of Lalande.

Mösting

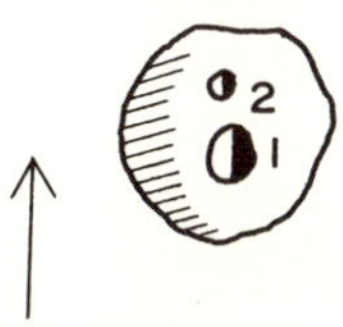

22. Müller (+036,-133;24/20)

(1) ⌞4/3; <0.1⌟ Very faint central elevation.

23. Murchison (-002,+089;58)

The west half of the floor is covered by a rugged plateau, and there is a heap on the north-east corner of the floor. Both regions are rich in mounds.

(1) ⌞7;0.5̇⌟ A mound.

(2) ⌞5.5;0.3⌟ A distinct peak.

(3) ⌞4/2;17⌟ A ridge with two peaks.

(4) ⌞5̇;0.14e⌟ (5) ⌞2.7;0.12e⌟ Ridges having easterly steep slopes.

(I)

(6) ⌞3;<0.2⌟ (7) ⌞2.5;<0.12⌟ (8) ⌞2;<0.1⌟ (9) ⌞1; <0.1⌟ Tiny mounds.

(10) ⌞3.5;<0.2⌟ A low, round hill.

24. <u>Palisa, U. F. to the E. of</u> (-099,-173;40)

(1) ⌞3.5/2.5;0.2⌟ A ridge-like peak. It seems as a true central peak.

(2) ⌞5/2.5;0.2⌟ (3) ⌞2.5/2.0;0.15⌟ (4) ⌞4.5/3;0.2⌟ Ridge-like peaks. Presumally, these are edges of the parallel valleys through this crater.

Murchison

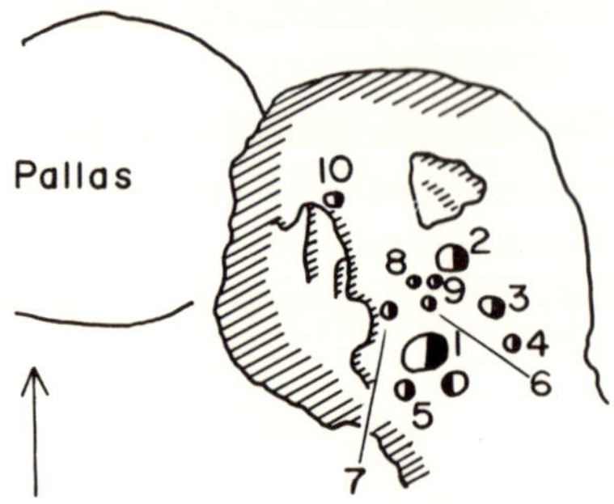

25. <u>Pallas</u> (-029,+096;50)

The floor is rich in tiny mounds, and very rugged.

(1) ⌞10;1.2⌟ A large central peak, having conical form. There are creases on the north and south sides.

(2) ⌞3.5;0.25e/0.4w⌟ (3) ⌞3;0.25⌟ Distinct mounds.

Pallas

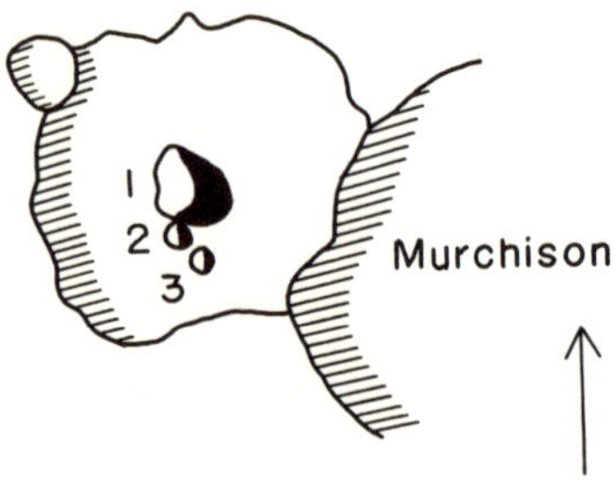

U.F. to the East of Palisa

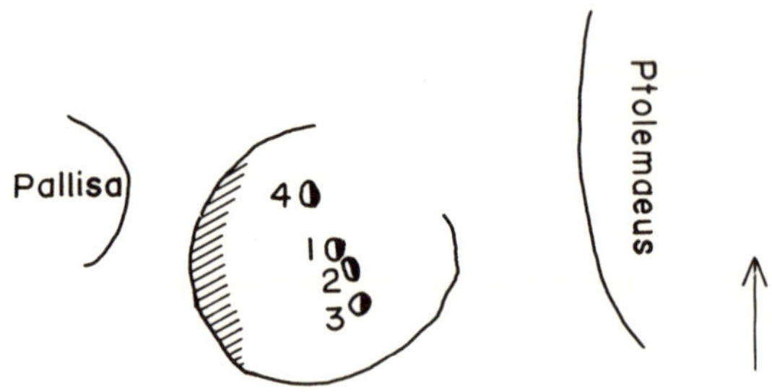

26. <u>Pallas F, U. F. to the S. of</u> (-019,+047;10)

(1) ⌞2;0.2⌟ Central peak, oblong in north-south direction.

27. <u>Pickering</u> (+122,-050;16)

(1) ⌞2;<0.1⌟ Very faint central elevation.

28. <u>Pickering, U. F. on the S.-W. of</u> (+197,-066;50)

(1) ⌞3;<0.1⌟ Very faint central elevation of the floor.

29. <u>Ptolemaeus, U. F. on the N.-E. of</u> (+012,-138;13/9)

(1) ⌞2.5;<0.1⌟ Very faint central elevation of the floor.

Rhäticus

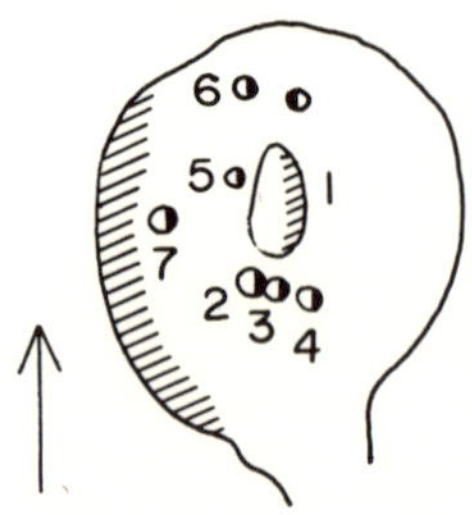

30. <u>Rhäticus</u> (+086,+000;49/43)

(1) ⌞11/8;<0.3⌟ A low central elevation, which looks like a remnant of twin mountains.

(2) ⌞ 2.5;<0.2⌟ A faint mound.

(3) ⌞1.5;<0.3⌟ (4) ⌞1.5;<0.3⌟ Faint peaks.

(5) ⌞1.5;<0.2⌟ A small peak on the west bottom of the (1).

(6) ⌞2.0; < 0.3⌟ A small mound on the bottom of the north wall.

(7) ⌞1.5;<0.3⌟ A rather distinct peak on the bottom of the west wall.

31. <u>Ritchey</u> (+145,-193;24)

(1) ⌞2.5;0.5⌟ A peak having somewhat ridge-like form, whose direction is parallel to the wrinkle structure around this crater.

32. <u>Saunder</u> (+152,-075;45)

(1) ⌞2.5;0.3⌟ The most distinct peak.

(2) ⌞1.5;<0.1⌟ (3) ⌞1.5;<0.1⌟ Small mounds.

(4) ⌞2.5;0.3⌟ Rather distinct mound.

(5) ⌞1;<0.1⌟ (6) ⌞5/2;<0.1⌟ Faint oblong hills.

(7) ⌞1.5;<0.1⌟ A low mound.

(8) ⌞1.5;<0.1⌟ A faint mound.

(9) ⌞1.5;0.1⌟ Rather distinct mound.

(10) ⌞1; U. L.⌟ An isolated minute peak.

Saunder

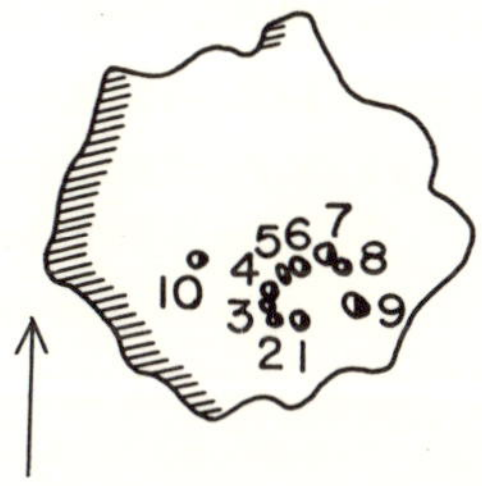

33. <u>Saunder, U. F. to the E. of</u> (+197,-066;50)

(1) ⌞20/8;1.8⌟

(2) ⌞12/11;1.2⌟

The central mountains are consist of two large blocks.

U.F. to the East of Saunder

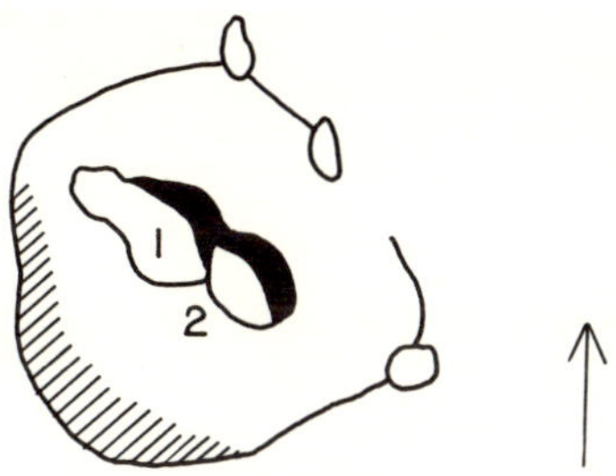

34. <u>Triesnecker</u> (+063, +073;26)

The floor has rhombic form, narrow and centrally elevated.

(1) ⌞2.5;0.35⌟ (2) ⌞3;0.5⌟ Two round domes.

(3) ⌞4;0.4⌟ A large round dome on slope of the south wall.

(4) ⌞3;<0.2⌟ (5) ⌞3;<0.2⌟ Twin domes on the north wall.

Triesnecker

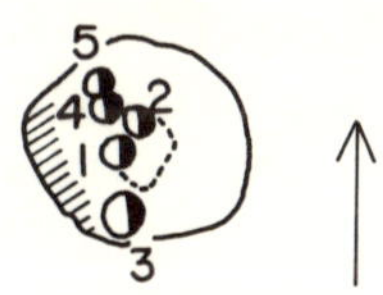

35. <u>Ukert</u> (+024,+134;24)

The floor is bowl-formed.

(1) ⌞4.5;0.9⌟ A central peak.

(2) ⌞3;<0.1⌟ A small mountain on the south of the central peak.

(3) ⌞2.5;<0.1⌟ A small mountain on the north of the central peak.

Ukert

II

1. Alfraganus (+324,-094;21)
   (1) ⌜7;U. L.⌝ Very weak central elevation. Its southern part is more elevated.
2. Alfraganus H (+327,-077;13/11)
   (1) ⌜3;<0.5̇⌝ A faint central hill on very shallow floor.
3. Anděl (+212,-181;34)
   (1) ⌜4;0.4⌝ The highest peak on the rugged heap, which covers almost half part of the floor.
   (2) ⌜2.7;0.2⌝ May be the true central peak.
4. Anděl, U. F. on the E. of (+228,-186;31/26)
   (1) ⌜5.5;0.6⌝ A large central peak.

Anděl

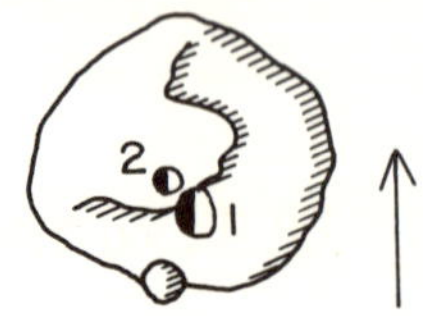

5. Arago (+363,+107;26)
   The floor is rather rugged.
   (1) ⌜4(width of the dyke);0.8⌝ From the somewhat oblong central peak, a long dyke runs to the north wall, of which north end is whitish. This dyke seems as a distinct part of the linear structure through this crater.

Arago

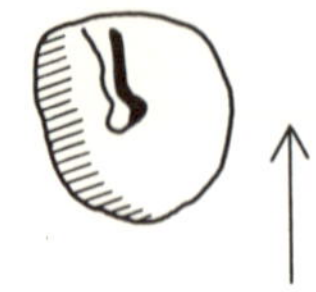

6. Capella (+567,-133;45)
   (1) ⌜11;2.1⌝ A large central hill. On the north side of its summit, an indistinct crease lies. On the south bottom, two hills are superposed.
7. Capella F (+572,-160;13)
   (1) ⌜2;<0.5⌝ A dyke reaches from south-east rampart to the central peak.
8. Capella F, U. F. on the S. of (+575,-168;15)
   (1) ⌜4;<0.2⌝ Very faintly centrally elevated floor.
9. Censorinus C (+561,-053;30)
   (1) ⌜3.5;1.0⌝ A round and distinct dome, standing on north-easterly eccentric position.
   (2) ⌜5.5;0.7⌝ A platform.
   (3) ⌜3.0;0.7⌝ A tiny peak. May be a true central peak.
   (4) ⌜1.5;<0.2⌝ A faint peak.
   (5) ⌜4;0.3⌝ A dome on the bottom of the south-east wall.
   (6) ⌜4/2;<0.3⌝ An oblong hill.
   (7) ⌜4;0.7⌝ A mountain having wide slope towards west.
   (8) ⌜12/5;1.2w⌝ From the south wall, long and lofty dyke runs to the peak (3)

Censorinus C

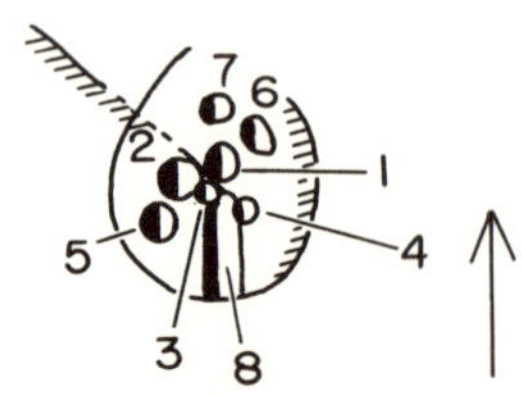

10. Censorinus E, U. F. to the N. of (+572,-085;30)
    This formation is a compound crater.
    (1) ⌜5̇;0.2⌝ An oblong mound.
11. Censorinus N (+596,-034;37)
    (1) ⌜5̇;0.4⌝ Conical-formed central peak.
12. Censorinus S (+589,-067;12)
    (1) ⌜5̇;<0.5̇⌝ At north-westerly eccentric position, there is an elevation which has wide slope towards west.
13. Delambre (+300,-034;53)
    (1) ⌜30;<0.5̇e⌝ (2) ⌜30;<0.5e⌝ Two scarps run through the floor, with easterly steep sides. Heights of the scarps are less than 0.5 km, according to the Ranger VIII photographs.

Delambre

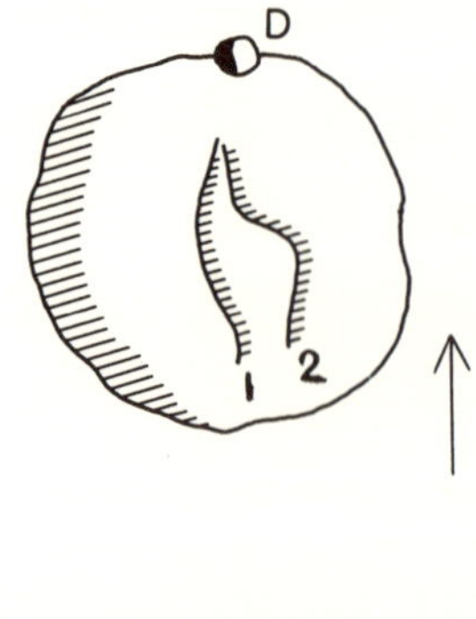

14. Dollond B, U. F. on the S. Rampart of (+232,-145;9)

(1) ⌊2;<0.2⌋ Very faint central mountain.

15. Dollond B, U. F. to the S. of (+231,-157;6)

(1) ⌊1;<0.1⌋ Very faint central elevation.

16. Dollond B, U. F. Containing (+264,-152;137)

(1) ⌊5.5;0.7e⌋ The highest peak.

(2) ⌊11;0.3⌋ (3) ⌊5;0.3⌋ These are likely to be central mountains of a obscure ring, whose rampart runs through the peak (1).

(4) ⌊5.5;0.3w⌋ This peak seems true central mountain of the present formation.

(5):α ⌊ 5.5;0.2⌋ This is probably a central peak of an obscure formation.

U.F. Containing the Dollond B

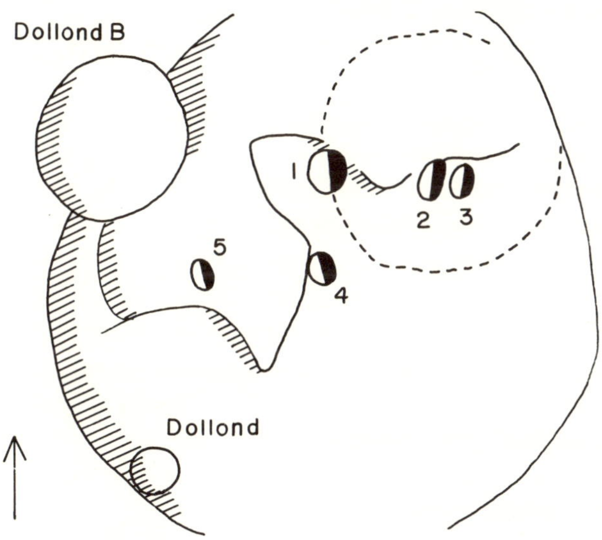

U.F. to the North-West of Dollond C

17. Dollond C, U. F. to the N.-W. of (+207,-107;10)

(1) ⌊2.5/2.0;0.3⌋ A ridge runs through south wall. The central peak belongs to another ridge, and looks like twin mountains which stand in a line parallel to the structure around the present crater.

18. Hypatia C (+355,-016;16)

(1) ⌊6/4;0.5⌋ An oblong central mound.

Isidorus B

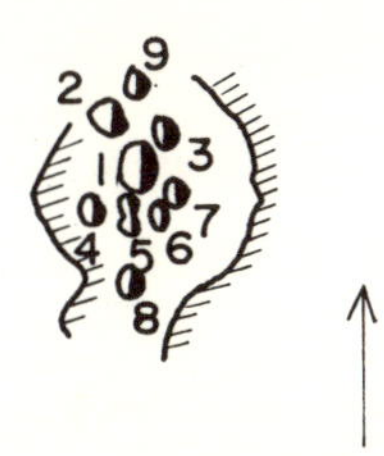

19. Isidorus B (+543,-078;30)

(1) ⌊5;<1.0⌋ A mound which is probably the true central peak.

(2) ⌊5;0.6⌋ The most dominant peak.

(3) ⌊6.5;1.0⌋ A rather broad peak.

(4) ⌊3;<0.35⌋ A low mound.

(5) ⌊5.5/2.7;<0.3w⌋ Twin mountains.

(6) ⌊2.8;<0.3⌋ (7) ⌊3.0;<0.3⌋ Low mounds.

(8) ⌊2.7;0.18w⌋ A dome standing at the neck of the south groove.

(9) ⌊3.0;<0.3⌋ The dome standing on slope of the north wall.

20. Jansen B (+441,+185;16) A compound crater.

(1) ⌊4;<0.2⌋ An elevation which is the limb of the accompanying crater.

(2) ⌊2.5;<0.2⌋ (3) ⌊2;<0.2⌋ Very weak elevations.

21. Julius Caeser, U. F. to the S. of (+274,+119;24) A mountain ring.

(1) ⌞3.5;0.1⌟ A distinct central peak.

Jansen B

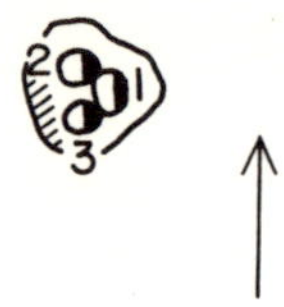

Kant

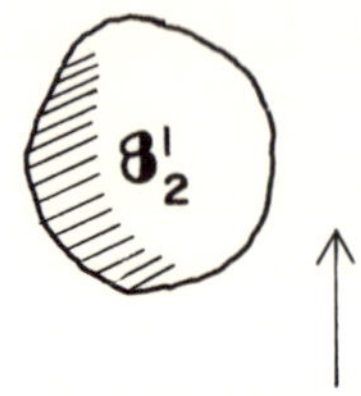

22. Kant (+340,-184;32)
    (1) ⌞4;0.8e⌟ (2) ⌞4;0.8e⌟ Central hill is twin, standing in north-south direction. The slopes are steeper easterly, and milder westerly.
23. Kant C (+370,-163;18)
    (1) ⌞9;<0.5̇⌟ (2) ⌞2;0.5̇⌟ The (1) is a large platform, of which north end is small peak (2).
24. Mädler (+487,-191;28)
    (1) ⌞about 4;<1⌟ Very low and broad central peak,,with gentle slope. It looks like a south end of a linear structure, which lies to north-east direction from this crater.

Kant C

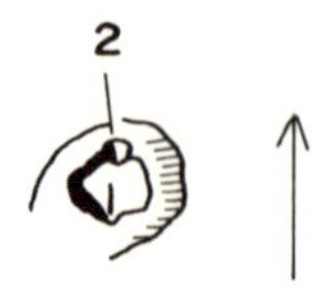

25. Maskelyne (+500,+038;24)
    (1) ⌞4;1⌟ At a little northerly eccentric position of the narrow and flat floor, a small central peak of cone form is seen. On its south-east bottom, there is a pit whose diameter is about 2 km.
26. Maskelyne A (+560,-001;29)
    (1) ⌞15;<0.6w/<0.4e⌟ A large mountain bolck. The highest part is a ridge on the north-east limb of the block.
27. Ritter (+329,+035;31)
    (1) ⌞1.4;<0.5̇⌟ (2) ⌞1.2;<0.4⌟ (3) ⌞2.4;<0.4⌟ On westerly eccentric position of the floor, three hills stand in north-south line. These hills have steep slope easterly, and form west bank of a crater chain, which is parallel to the other crater-chains around this crater, according to the Rahger VIII photographs.

Ritter

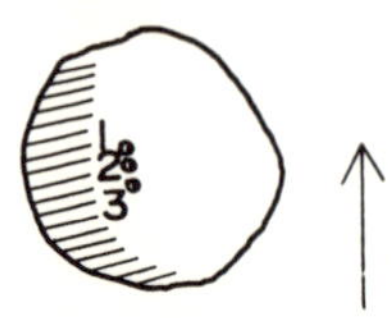

Sabine

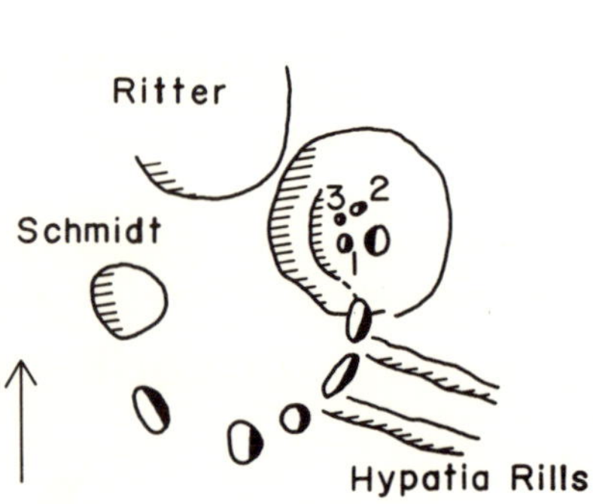

Ritter C

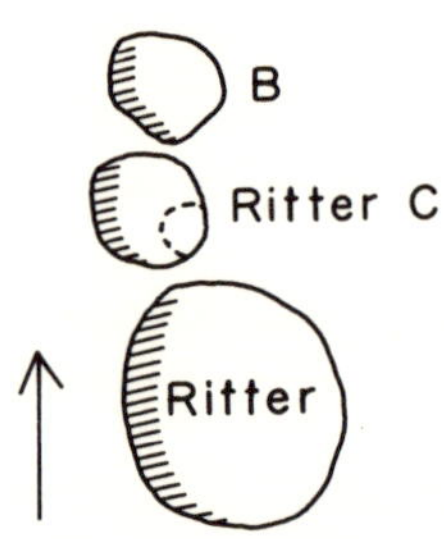

28. Ritter C (+323,+048; 14) A compound crater.
    (1) ⌞5.5;U. L.⌟ On the south-westerly eccentric part of the floor, there is a faint elevation.
29. Sabine (+343,+024;30) On the west of this formation, namely between Schmidt and Sabine, there is a ghost ring whose wall penetrates into the floor of Sabine.
    (1) ⌞6;<0.2⌟ (2) ⌞6/3;<0.2⌟ (3) ⌞2.5;<0.2⌟ There is a concentric structure on the floor, and these three hills form a part of this structure. The hill (2) has ridge-like form, according to the Ranger VIII photographs.
30. Silberschlag P (+206,+120;39/26)

(II)

(1) ⌞1.4;<0.2⌟ A small dome.

(2) ⌞2/1.5; 0.2⌟ A part of wrinkle ridge.

31. Silberschlag S (+208, +139;34/20)

(1) ⌞ 5.5/1.7;0.2⌟ A mountain. Probably the true central peak.

(2) ⌞8/3.5;<0.3⌟ (3) ⌞5.5/5.0;<0.2⌟ (4) ⌞8/2.5;<0.2⌟ These are three peaks which form a part of wrinkle structure around this formation.

(5) ⌞1.4;<0.2⌟ (6) ⌞1.5;<0.5⌟ Small peaks which are likely to be domes.

Silberschlag P

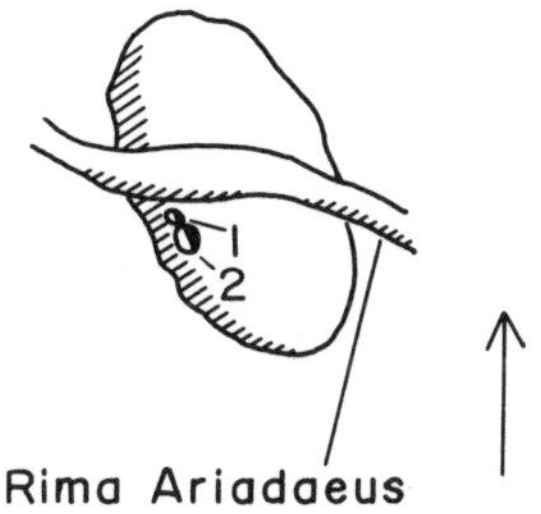

Silberschlag S

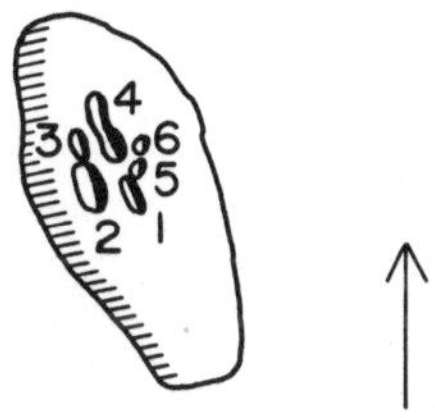

32. Taylor (+286,-094;41/34)

(1) ⌞9;1.4⌟ Conical-formed central peak.

(2) ⌞2.5;<0.5⌟ (3) ⌞1.5;<0.5⌟ These hills are promontries of the collapsed south wall.

Taylor

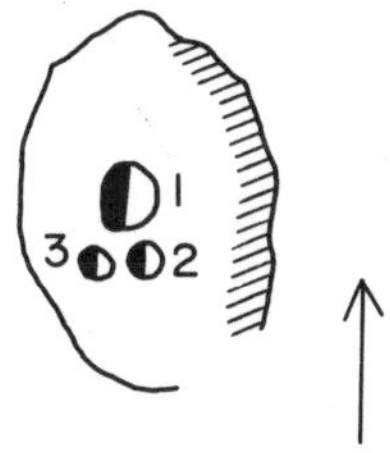

33. Taylor,U. F. on the N.-E. of (+301,-085;23)

(1) ⌞5;1.5w/0.5e⌟ A peak having conical form.

Theophilus

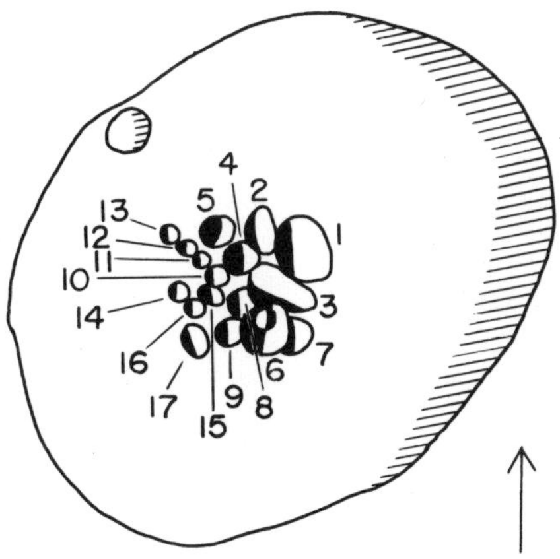

34. Theophilus (+435,-198;100)

(1):φ ⌞10;2.7⌟ The highest mountain.

(2):ψ ⌞5.5;1.1⌟ An oblong mountain.

(3):α ⌞10;2.1⌟ The largest hill.

(4) ⌞7;1.2⌟ A peak.

(5) ⌞5.5;0.7⌟ A low peak.

(6) ⌞7;0.7⌟ A round and broad hill, having a summit pit, whose diameter is about 2.7 km.

(7) ⌞4;0.5⌟ (8) ⌞4;<0.6⌟ (9) ⌞4;<0.5⌟ Low hills.

(10) ⌞2.8;<0.5⌟ (11) ⌞3.5;<0.5⌟ (12) ⌞3;<0.5⌟ (13) ⌞3;<0.5⌟ These four objects form a row of faint round mounds.

(14) ⌞4;<0.5⌟ (15) ⌞4;<0.5⌟ (16) ⌞3;<0.5⌟ Very weak mounds.

(17) ⌞5.5/3;<0.5⌟ An oblong hill.

35. Theophilus E (+404,-118;19)

(1) ⌞4 /2.7;<0.2⌟ (2) ⌞2.7;<0.2⌟ Very low and broad hills.

Theophilus E

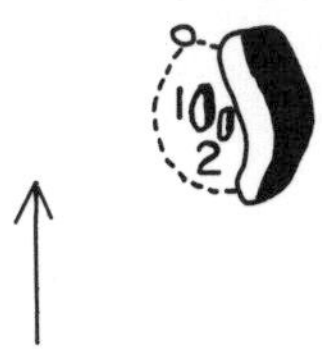

36. Theophilus F (+434,-139;12)

(1) ⌞4;<0.2⌟ A mound which is low, broad and oblong elevation.

(2) ⌞2.7;<0.2⌟ On south side of (1), small and whitish peak is attached.

(II)

37. Torricelli M (+517,-064;14)
    (1) ⌞3;<0.2⌟ Very faint central elevation of the floor, on northerly eccentric position.
38. Zöllner (+321,-140;47/36)
    (1) ⌞7.5;1.2⌟ The main peak.
    (2) ⌞4;0.8⌟ A small hill on the north bottom of the main peak.
    (3) ⌞5;1.3w⌟ (4) ⌞8;1.6⌟ Hills which may be collapsed wall.

Theophilus F

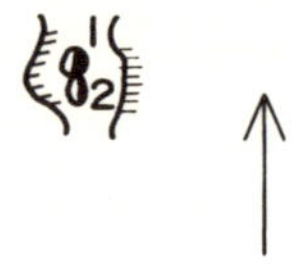

Zöllner

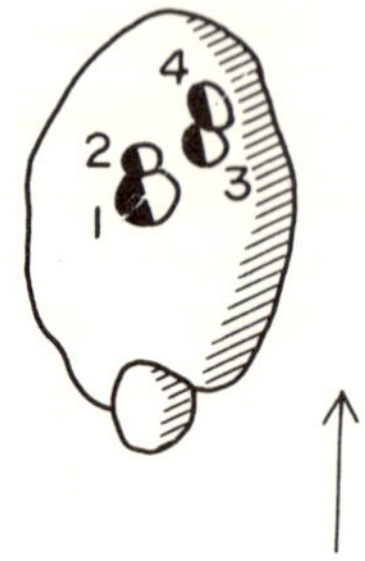

39. Zöllner, U. F. to the W. of (+299,-150;19)
    This is a crater of very irregular form. Probably superposition of two or more craters.
    (1) ⌞7/5.5;0.7⌟ The central hill of ridge-like form, lying in north-south direction.
40. Dimple Crater in Mare Tranquillitatis (+4158,+0473;0.031)
    (1) ⌞0.022;- -⌟ (2) ⌞0.013;- - ⌟ (3) ⌞0.010;- - ⌟ A crater whose form is somewhat deviated from perfect circle. Three central bodies lie on a line of north-south direction, according to the Ranger VIII photograph B-90.
41. Dimple Crater in Mare Tranquillitatis (+4150,+0464;0.098)
    (1) ⌞0.004;- - ⌟ This crater is found in the Ranger VIII photographs P-19 of the P-4 camera.

Dimple Crater (B 90)

Dimple Crater (P 19)

U.F. on the East of Jansen

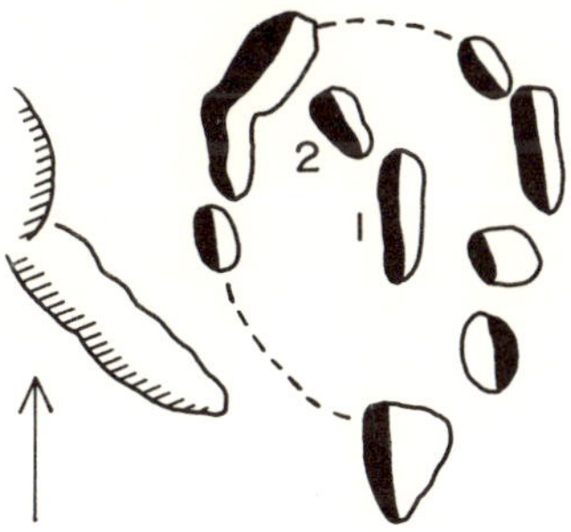

1. Bond, G. (+500,+536;20)
   (1) ⌞6;U. L.⌟ Small, centrally elevated floor.
2. Jansen, U. F. on the E. of (+494,+234;42)
   (1) ⌞12/4;<0.2⌟ This is a ridge which has easterly steep slope and forms east limb of a broad platform, whose diameter is about 14 km.
   (2) ⌞1$\dot{0}$/6;<0.2⌟ A ridge having westerly steep slope. These two ridges are part of wrinkle structure through this crater.

Menelaus

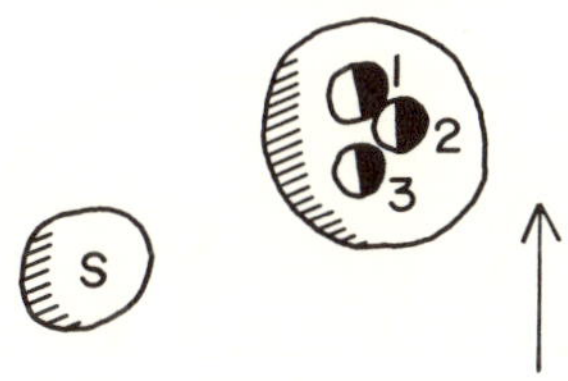

3. Menelaus (+264,+280;27) A ray crater.
   (1) ⌞4;<0.2⌟ (2) ⌞3;<0.2⌟ (3) ⌞4;<0.2⌟ Round and low hills.
4. Menelaus S (+246,+271;14)
   (1) ⌞1.5;<0.2⌟ Very tiny central hill.

Newcomb

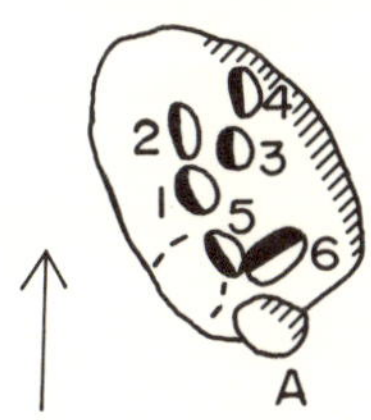

5. Newcomb (+600,+498;39)
   (1) ⌞4;1.0⌟ The central hill standing on westerly eccentric position of the floor.
   (2) ⌞4;U. L.⌟ (3) ⌞$\dot{5}$;U. L.⌟ Attached mounds of the central hill. hill. The (1) and (2) are whitish.
   (4) ⌞4;U. L.⌟ A hill which is probably a part of terrace structure of the north-west wall.
   (5) ⌞4;2w⌟ Probably a part of wall of an unnamed shallow crater.
   (6) ⌞1$\dot{0}$/4;1.5w⌟ A long dyke which is probably a part of the terrace structure of the south wall.

Plinius

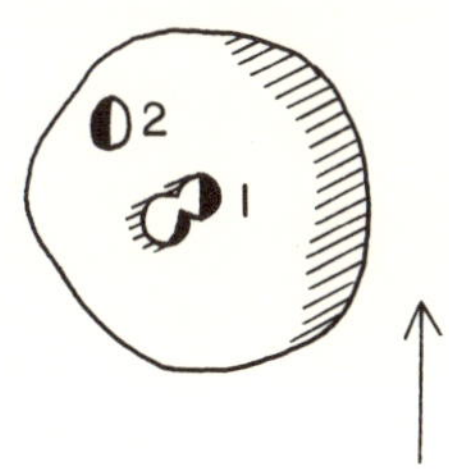

6. Plinius (+387,+265;43)
   (1) ⌞9;0.9⌟ The central mountain is composed of twin craters. One is on the center of the floor, and the other is its south-west neighbour. Diameter of the former one is about 3 km, and that of the last one is 5 km.
   (2) ⌞4;<0.7⌟ A rugged mountain at the bottom of the north-west rampart.
7. Posidonius (+424,+528;101)
   (1) ⌞5.5;0.8⌟ (2) ⌞2;0.3⌟ (3) ⌞3;0.3⌟ (4) ⌞4;0.4⌟ These hills stand as a mountain ring. Probably remnant of a old ring.
8. Römer (+537,+429;40)
   (1) ⌞14;0.9⌟ A large conical central mountain. According to Wilkins and Moore, it bears a summit craterlet.
9. Ross (+363,+202;26)

Posidonius

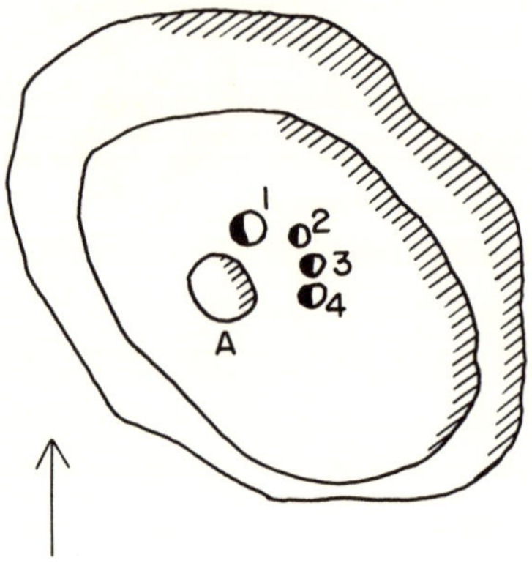

(III)

(1) ⌊3;<0.1⌋ A very slight elevation of the floor.

(2) ⌊1.2;<0.1⌋ A tiny hill.

Ross

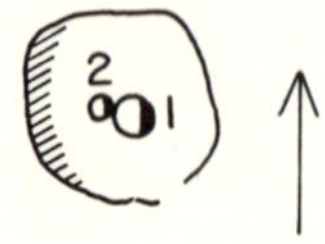

IV

Aristillus

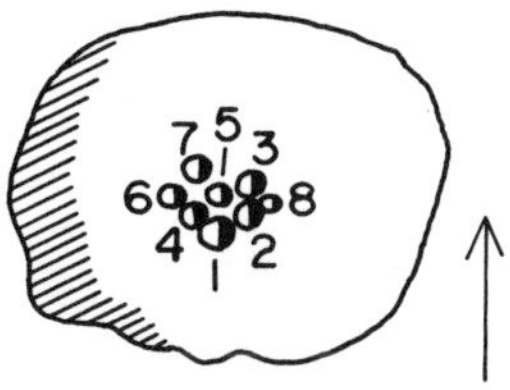

1. Aristillus (+018,+557;55)

(1) ⌞3.8;0.6e⌟ (2) ⌞2.7;0.8e⌟ (3) ⌞3.8; 1.2e⌟ (4) ⌞3.2;0.3(from east valley)⌟ (5) ⌞3.0;0.3(from east valley)⌟ (6) ⌞2.5; 0.4e⌟ (7) ⌞3.0;0.5e⌟ These are multiple peaks arranged in triple lines.

(8) ⌞1.4;0.2e⌟ A very faint mound on the east bottom of the central mountains.

Autolycus

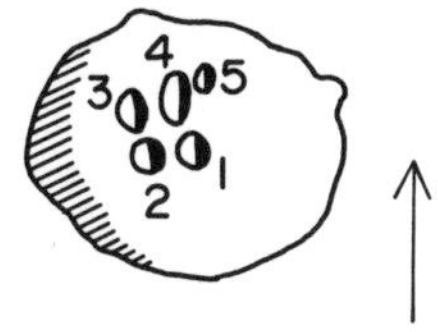

2. Autolycus (+022,+557;39)

(1) ⌞2.5;<0.15⌟ Very tiny peak, but it is likely to be the true central peak.

(2) ⌞2.7;<0.15⌟ (3) ⌞4.0/3.5;<0.15⌟ (4) ⌞5.5/3.5;<0.15⌟ These are ridges which constitute a part of the linear structure on the floor.

(5) ⌞2;<0.15⌟ A very faint ridge.

3. Eratosthenes (-190,+250;58) The floor is very rugged

Eratosthenes

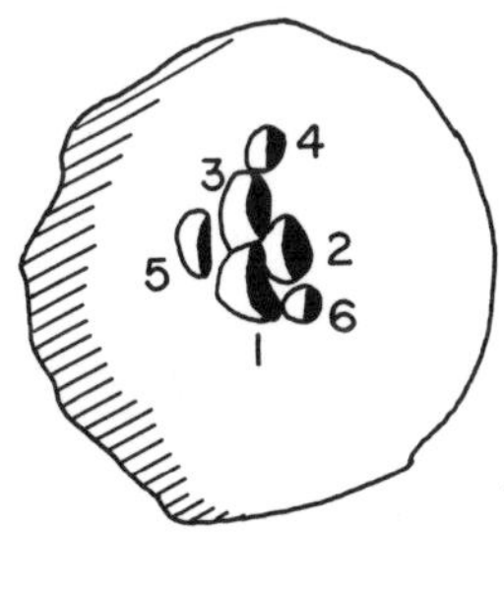

(1) ⌞8/7;1.2⌟ A mountain which may be the true central peak.

(2) ⌞5.5;1.1⌟ (3) ⌞7/5.5;0.9⌟ (4) ⌞4.0/2.7;0.8e⌟ Ridges which form a part of wall remnant on the floor.

(5) ⌞7.0̇/2.7;<0.1e⌟ Faint ridge standing on the west bottom of the (3).

(6) ⌞5̇;U. L.⌟ Very faint elevation.

Manilius

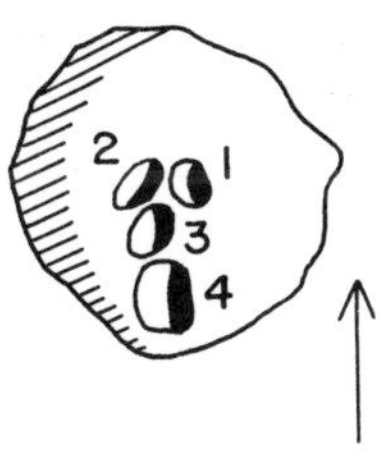

4. Manilius (+153,+250;39)

(1) ⌞3.5;0.5e⌟ (2) ⌞4;0.9w⌟ (3) ⌞4/3.5; 0.15⌟ These three peaks are likely to be a trified central mountain, each has steep slope outwardly.

(4) ⌞7;0.6⌟ A rugged ridge through the south wall.

V

Delisle

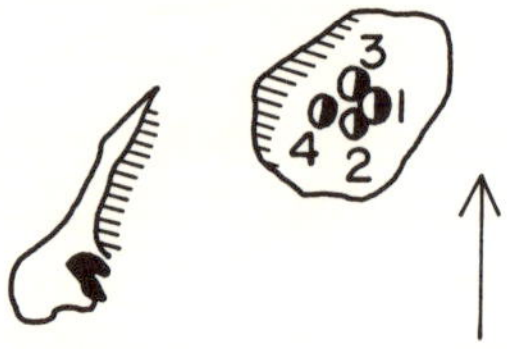

1. Delisle (-492,+500;25)

   (1) ⌞2.5;<0.15⌟ (2) ⌞2.5;<0.15⌟ (3) ⌞2.5;<0.15⌟ (4) ⌞2.5;<0.15⌟ Four faint mounds on rather rugged, centrally swelling floor.

2. Diophantus (-499,+463;19) This crater looks like compound crater.

   (1) ⌞5.5;<0.2⌟ Faintly swelling floor.

3. Euler (-447,+395;27)

   (1) ⌞10;<1.4⌟ Very large, but low central mountain.

Gay-Lussac A

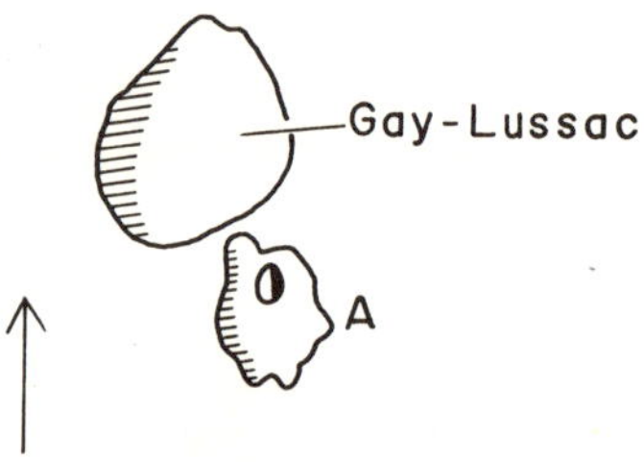

4. Gay-Lussac A (-339,+228;14)

   (1) ⌞4;0.5⌟ On northerly eecentric position, there is a tiny hill. This is probably wall remnant of a superposed crater.

Lambert

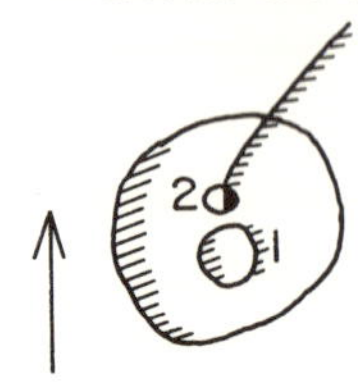

5. Lambert (-322,+435;30)

   (1) ⌞10;0.8⌟ This is a floor-crater on southerly eccentric position. The highest part of its wall is north-east.

   (2) ⌞3.1;0.6⌟ A hill, which is an end of a linearment passing through the north wall.

T. Mayer

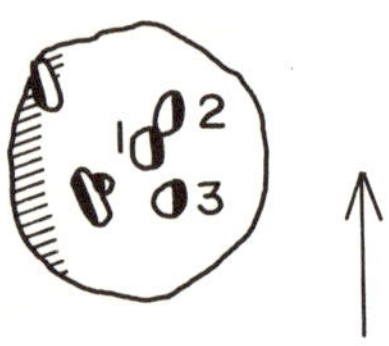

6. Mayer, T. (-469,+268;33) On south-westerly eccentric position, there is a crater chain.

   (1) ⌞3;0.2⌟ The central hill on the north-easterly eccentric position.

   (2) ⌞4/3;<0.2⌟ (3) ⌞3;<0.2⌟ Very low ridges.

7. Mayer, T. A (-457,+263;16)

   (1) ⌞3;0.8⌟ A very low hill on westerly eccentric position.

8. Mayer, T. C (-428,+212;16)

   (1) ⌞5.5;1.0(from its center)⌟ A double crater, like Hesiodus A.

Pytheas

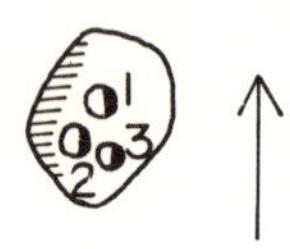

9. Pytheas (-329,+351;20)

   (1) ⌞4;0.9⌟ A low hill, which is the central peak.

   (2) ⌞4;0.6⌟ (3) ⌞3.5;0.6⌟ Two hills on the south side of the (1).

10. Timocharis (-202,+449;35)

   (1) ⌞5;1.2⌟ A central crater, whose highest part is east wall. On its south side, there is a narrow flat floor.

VI

1. Copernicus (-337,+168;93) The southern part of the floor is rather rugged, but the northern one looks very smooth.
   (1) ⌞9/5̇;1.0⌟ A mountain separated into two parts by a faint crease.
   (2) ⌞3;0.6⌟ A small hill, attached on the north end of the (1).
   (3) ⌞3;0.7⌟ A hill standing at the center of the floor. It is presumably the true central peak.
   (4) ⌞7.5/5.5;0.9⌟ A large mountain block.
   (5) ⌞1.5;0.3⌟ (6) ⌞4/3;0.3⌟ Low hills.
   (7) ⌞3;<0.2⌟ Very faint elevation.
   (8) ⌞1̇0;<0.2⌟ A V-formed low hill.
   (9) ⌞5.5;<0.2⌟ (10) ⌞4;<0.2⌟ Very faint mounds.

Copernicus

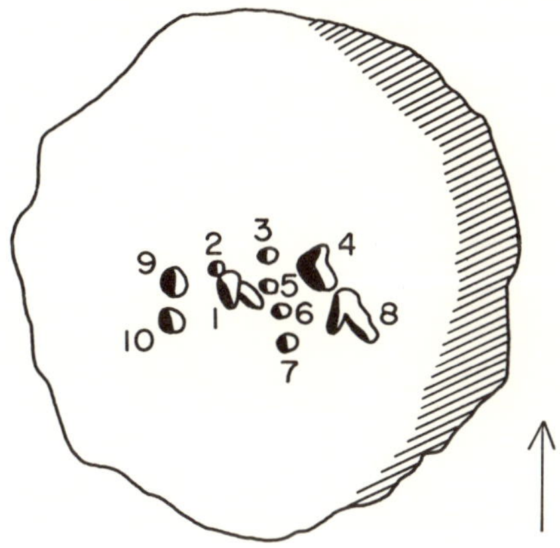

2. Encke (-595,+080;29)
   (1) ⌞2.7;<0.3⌟ (2) ⌞2.5̇;<0.3⌟ (3) ⌞2.7;<0.3⌟ Very low mountains, lying on a north-south line.

Encke

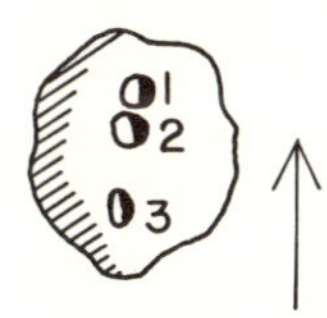

3. Guericke (-239,-200;58)
   (1) ⌞5̇;0.5⌟ (2) ⌞3.5;0.3⌟ (3) ⌞4;0.3⌟ Hills which are part of the linear structure around this crater.
   (4) ⌞5̇;0.4⌟ A faint, dome-like elevation.
   (5) ⌞1.5;0.2⌟ (6) ⌞1.5;0.2⌟ Faint small domes, according to the Ranger VII photographs.

Guericke

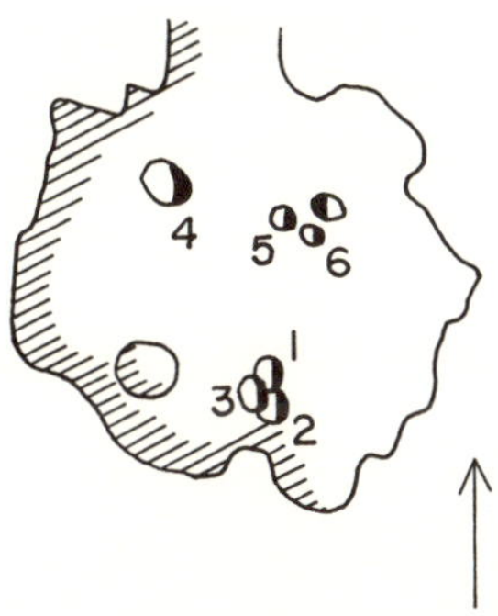

4. Hortensius (-466,+113;15)
   Depth of the floor is about 3 km.
   (1) ⌞1.6;<0.5⌟ A small mound on the southerly eccentric position.
5. Kunowsky (-536,+056;18)
   (1) ⌞3.5;<0.2⌟ A very gentle mound on the center of smooth floor.
6. Lansberg (-448,-006;40)
   (1) ⌞6;1.3⌟ (2) ⌞6;1.2⌟ A large central mountain, whose diameter is about 10km, with two peaks arranged on north-south line. The forms of both peaks are not conic.

Lansberg

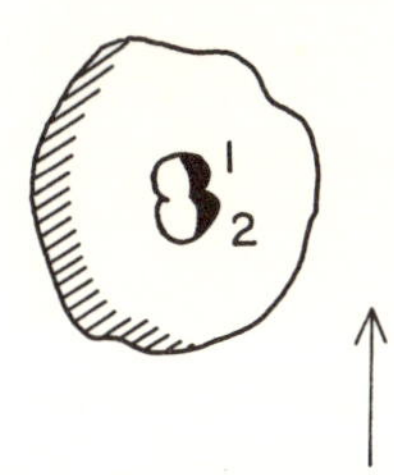

7. Reinhold (-387,+057;47)
   (1) ⌞3;0.9⌟ The true central peak.
   (2) ⌞2.5;0.7⌟ A small hill to the south of the (1).
   (3) ⌞4;1.0⌟ A hill which is a part of linear structure running through the north wall.
   (4) ⌞2.5;1.0⌟ A hill standing on the east limb of the

Reinhold

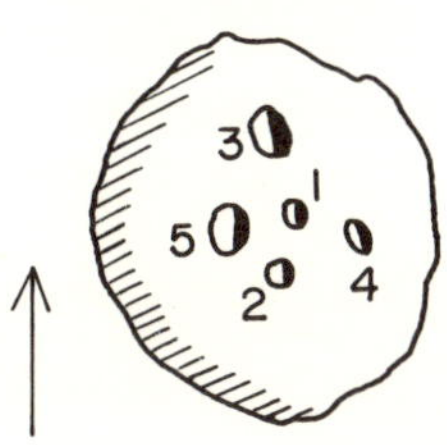

floor, which looks like a part of collapsed wall. Around the hill (4), there are several small objects which look like remnants of collapsed wall on the bottom of the wall of the present crater.

(5) ⌜5;U. L.⌝ A hill which is a elevated part of the floor.

VII

1. Agatharchides (-483,-338;49)
The central mountain looks like a white patch at the time of very low illumination.
(1) ⌞5.5;0.5̇⌟ (2) ⌞3.5;0.2⌟ There are large twin mountains on the bottom of east wall.

Agatharchides

2. Bullialdus (-353,-354;59)
Multiple central peaks.
(1) ⌞4;lw⌟ (2) ⌞3.5;lw⌟ These two peaks have westerly long slopes.
(3) ⌞3.5;1.5̇e⌟ A peak having easterly long slope.
(4) ⌞2.5;0.6e⌟ (5) ⌞1.7;0.4e⌟ Peak (5) is separated from the peaks (3) and (4) by faint creases.

Bullialdus

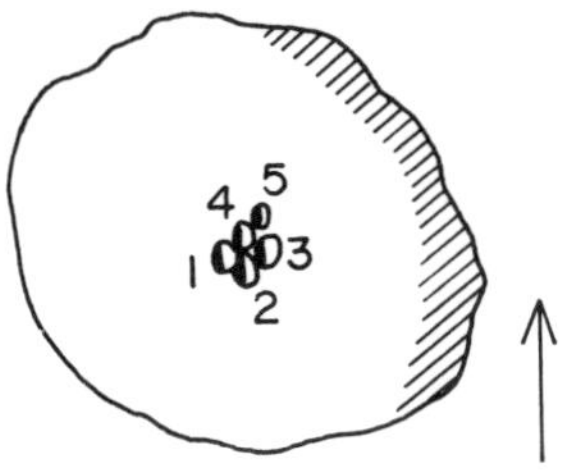

3. Bullialdus B (-342,-397;21)
(1) ⌞7/3.5;0.3⌟ An oblong formation running from the north wall.

4. Campanus (-411,-469;48)
(1) ⌞5̇/4;0.5̇w⌟ (2) ⌞3.5;0.3e⌟ Two hills stand in uncinate form.

Campanus

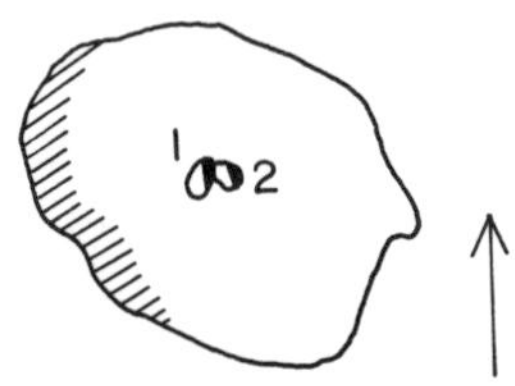

5. Cichus (-302,-549;39)
The floor is not smooth.
(1) ⌞3.5;0.6⌟ Rather prominent peak. Probably the true central peak.
(2) ⌞3.5;0.4⌟ (3) ⌞4;0.4⌟ Faint hills.

Cichus

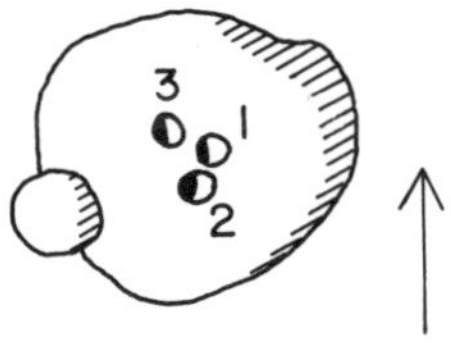

6. Cichus G, U. F. to the S. of (-332,-596;40)
In the Arizona Catalogue, this formation is an arrgegate of three craters, No. 33539 (ξ=-0.330, η=-0.596), No.33539A (ξ=-0.336, η=-0.594), and No. 33539B (ξ=-0.338, η=-0.594).
(1) ⌞5/3;0.4⌟ (2) ⌞2;0.3⌟ These hills are remnant of collapsed wall of No.33539A.
(2) ⌞2.5;0.4⌟ The central hill of No.33539A.

U.F. to the South of Cichus G

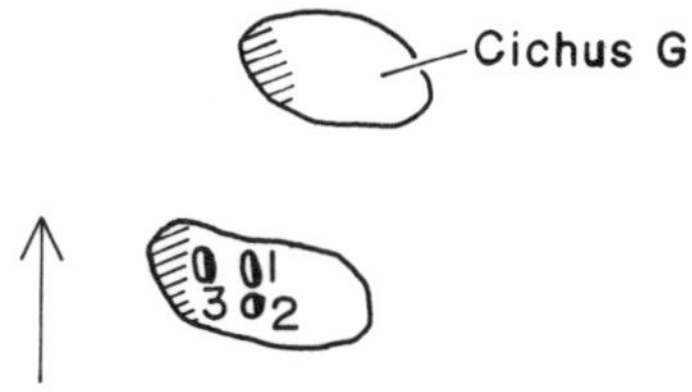

7. Clausius C (-512,-580;15)
(1) ⌞3;U. L.⌟ Faintly elevated floor.

8. Clausius F, U. F. around (-497,-595;61)
(1) ⌞26;0.5⌟ This is a double crater, and Clausius F is the inner one.

9. Darney (-386,-252;15)
(1) ⌞2.5;<0.2⌟ A tiny central mound.

10. Doppelmayer (-581,-477;64)
(1) ⌞14/8;0.7⌟ A large and oblong central mountain.

(VII)

(2) ⌜10;0.4⌝ (3) ⌜10̇/6;0.3⌝ (4) ⌜8/4;0.15⌝ (5) ⌜6/4;<0.15⌝ Floor hills which have form of ridge and are all parallel.

Doppelmayer

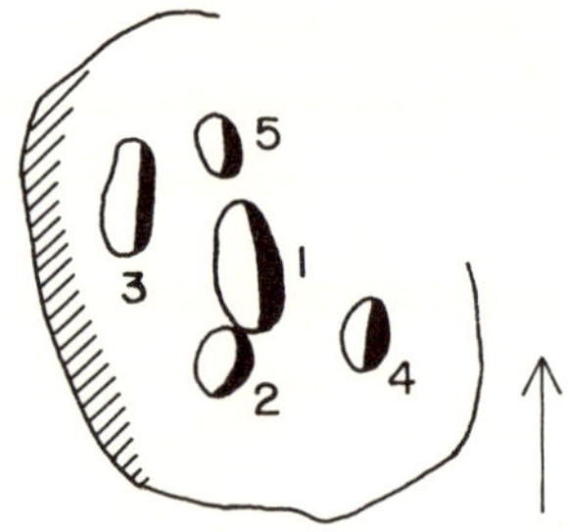

11. Elger (-405,-578;21)
    (1) ⌜3;0.3⌝ A tiny hill on the southerly eccentric position.
12. Hesiodus (-245,-491;42)
    (1) ⌜5̇;U. L.⌝ A typical central pit, which is named as Hesiodus D.
13. Hesiodus, U. F. to the N. of (-250,-472;27)
    (1) ⌜3.5;0.3⌝ A central mound having a slightly oblong form in direction from north-east to south-west.
14. Hesiodus A (-253,-501;15)
    (1) ⌜7.5;0.3⌝ A typical concentric crater.
    (2) ⌜2.5/1.5; U. L.⌝ An oblong, minute ridge.
    (3) ⌜0.5; U. L.⌝ A very minute mound.

Hesiodus A

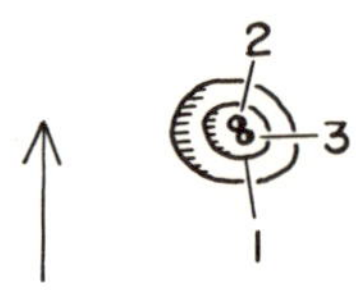

15. König (-380,-410;22) The floor is very rugged.
    (1) ⌜6/4;0.3⌝ (2) ⌜4;0.3⌝ These hills may be collapsed part of west wall.
    (3) ⌜4/1.5;0.2⌝ Very fine ridge at the northerly eccentric position.

König

Pitatus

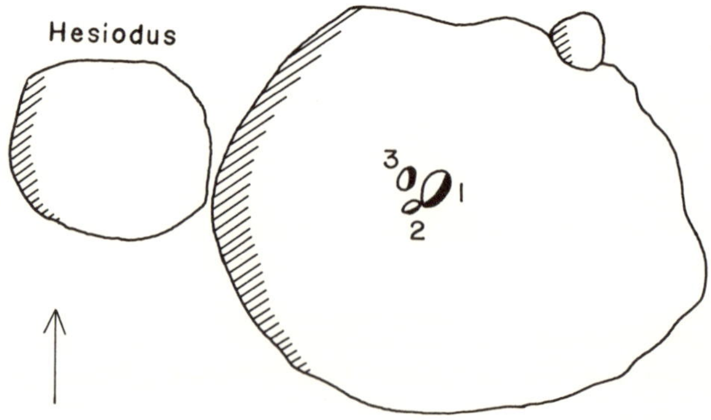

16. Pitatus (-203,-497;60) Near periphery of the crater plane, there are several hills which are likely to be remnants of collapsed wall.
    (1):α ⌜11/5̇;1.0⌝ The central mountain standing on westerly eccentric position of the floor.
    (2) ⌜4/2;0.5⌝ A small hill attached on the south-west side of the (1), making uncinate form with it.
    (3) ⌜2.5;<0.2⌝ Very faint mound.
17. Pitatus H (-232,-508;16)
    (1) ⌜2;U. L.⌝ A central hill standing on somewhat easterly eccentric position.
18. Puiseux (-557,-466;25)
    (1) ⌜4;0.1⌝ Very low, but broad central hill.

19. Vitello (-525,-506;45)
    (1) [6;0.7] A central peak of conical form on rugged floor.

20. Wurzelbauer (-228,-558;82)
    (1) [4;1e] This crater looks like a superposition of two old craters. At the center of the outer crater, there is a small mound which is likely to be a central hill. The floor is very rugged.

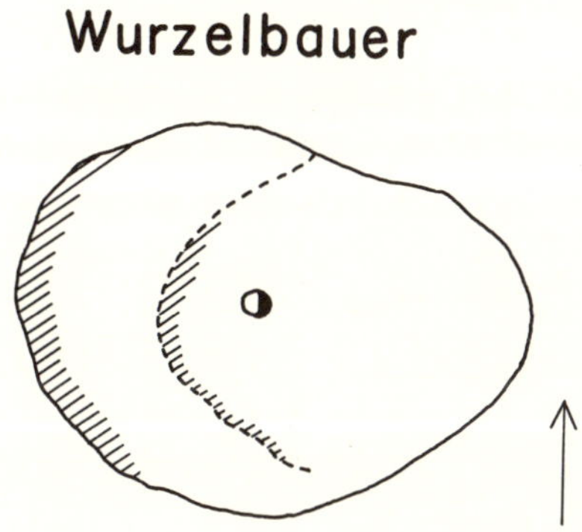

VIII

1. Abenezra (+194,-358;42)

(1) ⌞5.5;1.0̇⌟ The central peak, having conical form.

(2) ⌞4.0;0.3⌟ A mountain, which is a part of oblong formation on slope of the north wall.

(3) ⌞4.5;0.3⌟ A mound having a mountain-like form.

(4) ⌞1̇0/3;0.2⌟ A long curved fault, which looks like a lava flow.

Abenezra

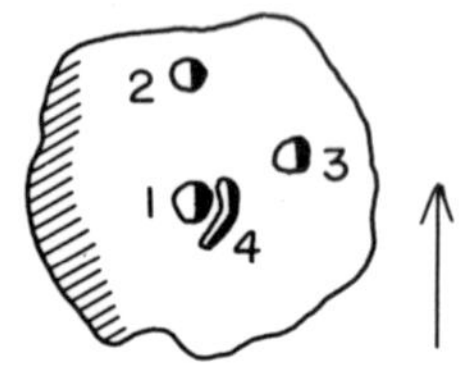

2. Abenezra C (+179,-364;41)

(1) ⌞8;2⌟ This crater is superposed by Abenezra, and the central mountain may be a part of its west wall.

3. Abulfeda C (+184,-221;17)

(1) ⌞2.5;0.3⌟ (2) ⌞2.5;0.2⌟ Both peaks are oblong and parallel to each other, and also parallel to the linear structure around this crater.

Abulfeda C

4. Airy (+094,-311;37)

(1) ⌞7;1.1e/1.3w⌟ An eminent central mountain, having a conical form. To the north and south of it, there are two ridges, parallel to each other, which are part of the linear structure around the present crater.

5. Airy, U. F. to the E. of (+159,-304;25)

(1) ⌞2.5;U. L.⌟ Very faint central mound.

6. Airy B (+141,-303;29)

(1) ⌞8.5/4.0;0.7⌟ A ridge-like central peak on northerly eccentric position. Probably a part of linear structure through the present crater.

Airy T

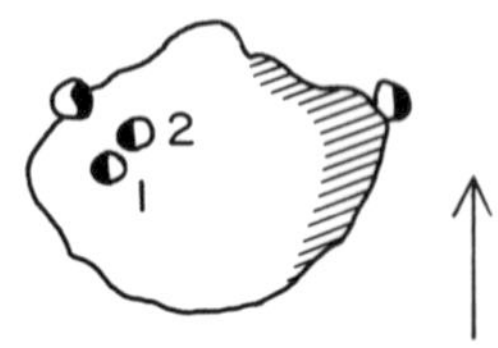

7. Airy T (+155,-329;40)

(1) ⌞4;0.8⌟ (2) ⌞3;0.4⌟ Two mountains stand at the bottom of the north-west wall. Both mountains are not conic.

8. Aliacensis (+078,-510;80)

The floor is very flat.

(1) ⌞5̇;0.8⌟ A cone-formed central peak.

(2) ⌞4/2;<0.2⌟ (3) ⌞2;<0.2⌟ Faint ridges.

Aliacensis

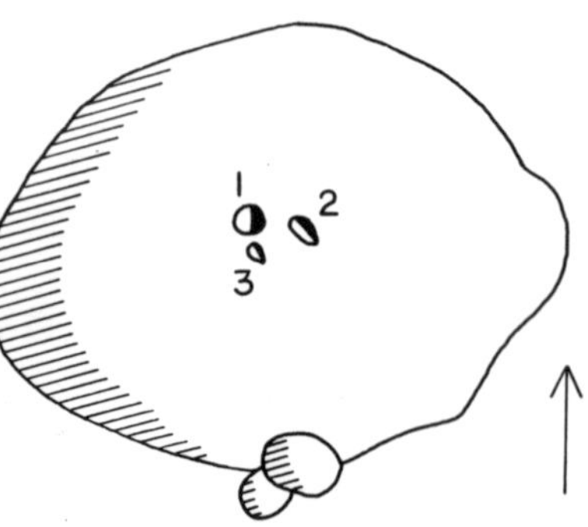

9. Aliacensis, U. F. to the E. of (+122,-501;5)

(1) ⌞1;<0.1⌟ Very tiny central object.

10. Aliacensis, U. F. to the S.-E. of (+096,-543;7)

(1) ⌞1.2;U. L.⌟ Very faint central elevation.

11. Aliacensis A (+113,-495;14)

(1) ⌞1;<0.1⌟ A faint central peak.

12. Aliacensis B (+047,-520;19)

(1) ⌞7;U. L.⌟ Very broad, centrally elevated floor.

13. Alpetragius (-075,-276;40)

(1) ⌜12.3;2.0⌟ Very large central mountain. There is a summit pit whose diameter is 1.6 km, according to the Ranger IX photographs.

(2) ⌜11;0.3⌟ On slope of the north wall, there is a broad hill.

Alpetragius

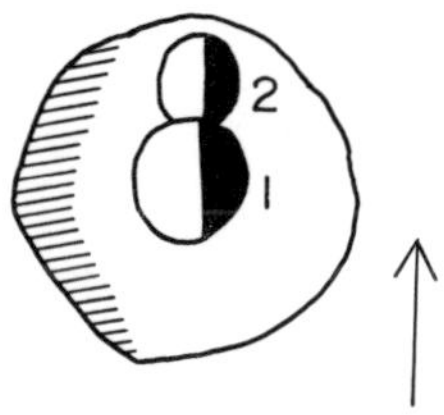

14. Alphonsus (-046,-233;119)

Alphonsus

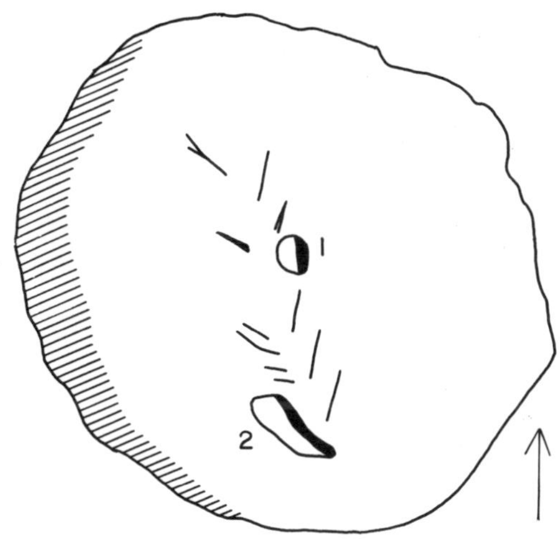

(1):*α* ⌜6.6;2.21⌟ The central peak having very steep slope. It is somewhat oblong along the direction of the local linear structure. According to the Ranger IX photographs, there is a summit pit whose major and minor diameter is 0.44 km and 0.18 km respectively. The peak is very whitish.

(2) ⌜25/5;0.3⌟ Through west side of the main peak, a group of low ridges runs. These ridges form a lattice structure, being composed of two components. The (2) is the most eminent one near the south wall.

15. Alphonsus B (-003,-229;24)

The wall and floor of this crater are very disturbed.

(1) ⌜2.3;0.5⌟ A minute central peak.

16. Apianus C, U. F. to the N. of (+158,-465;5)

(1) ⌜0.7;U. L.⌟ Very minute mound.

Argelander

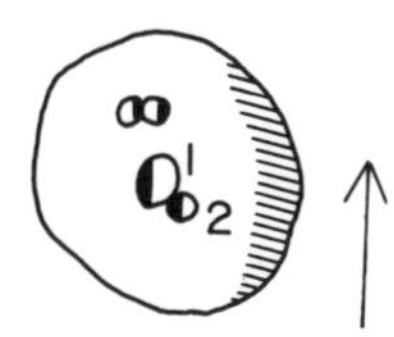

17. Argelander (+097,-285;34)

There are twin floor craters on the northern part of the floor.

(1) ⌜6;0.8⌟ A comparatively eminent central peak.

(2) ⌜4;0.4⌟ Smaller hill attached on the south-east bottom of the (1).

18. Arzachel (-031,-313;97)

There is a comparatively large floor crater, whose diameter is about 10 km, on the easterly eccentric position of the floor. Several smaller floor craters can be perceived. A distinct crack runs along the east wall.

(1):*γ* ⌜13.4;2.1⌟ Principal peak standing on the westerly eccentric position of the floor. The crease on the south bottom looks like a small crater. In the photographs of Ranger IX, we cannot find any summit pit on it.

(2) ⌜$\dot{5}$;0.6⌟ (3) ⌜$\dot{5}$;0.4⌟ Twin mountains, which may be extensions of the crease on the south bottom of the main peak.

(4) ⌜2.6;0.3⌟ A small mountain, attached to the north-east bottom of the main

peak.

(5) ⌞2.0;0.2⌟ (6) ⌞1.0;0.1⌟ (7) ⌞1.0;0.1⌟ (8) ⌞1.5;0.15⌟ Minute mounds on the floor.

(9) ⌞2.0;0.2⌟ (10) ⌞4/3;0.2⌟ Broad hills.

(11) ⌞1;U. L.⌟ (12) ⌞1.5;U. L.⌟ (13) ⌞1;U. L.⌟ Very faint mounds.

(14) ⌞2.5/1.5;0.2⌟ (15) ⌞5/3;0.3⌟ Oblong hills.

(16) ⌞4;0.3⌟ (17) ⌞7;0.2⌟ Broad plateaus.

Arzachel

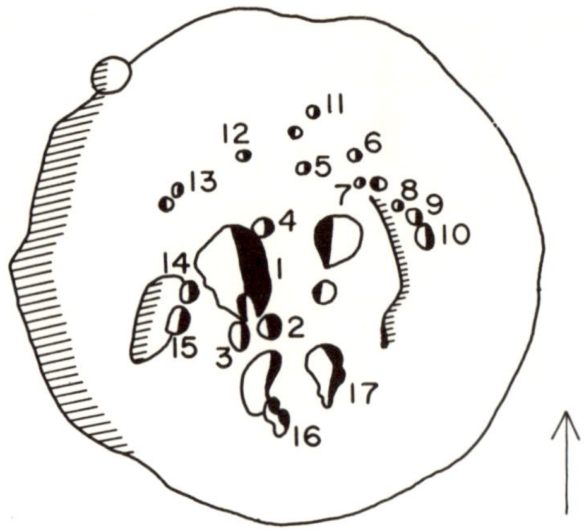

19. Azophi A (+178,-413;29)

(1) ⌞6;0.2⌟ On the south part of the floor, there is a large mountain block, having a summit crater, whose diameter is about 2.7 km. This summit crater is named Azophi AA.

20. Azophi B (+170,-400;19)

(1) ⌞1.4;0.1⌟ Very minute central peak.

Ball A

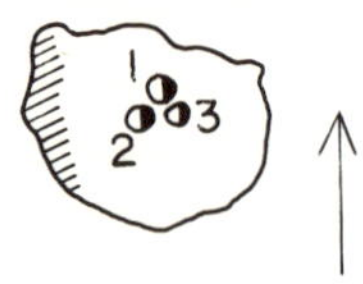

21. Azophi G (+195,-405;53)

This formation is also called as Sesibuto. Northern part of the rugged and rich with small pits.

(1) ⌞10;0.7⌟ A large mountain block on the south-westerly eccentric position of the floor, whose form is faintly twin-like.

22. Ball (-118,-587;40)

(1) ⌞15̇;1.5̇⌟ A large central mountain. On its south slope, there are two craters. Diameter of the larger one is 7 km, and that of the smaller one is 1.5 km. A large dislocation runs from the south wall to the north part of the floor.

Burnham

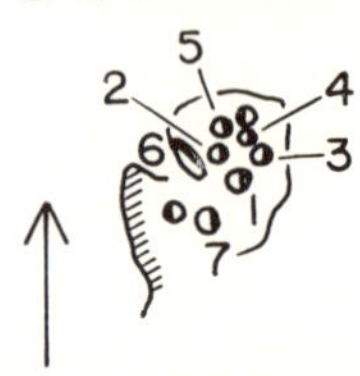

23. Ball A (-133,-570;29)

(1) ⌞2.5;<0.2⌟ (2) ⌞2.5;<0.2⌟ (3) ⌞2.5;<0.2⌟. Low triple mounds on the floor.

24. Blanchinus, U. F. on the W. Wall of (+029,-434;9)

(1) ⌞0.4;U. L.⌟ On occasion of a special illumination, a small white patch can be seen on a little darker floor.

25. Burnham (+124,-240;22) Extremely damaged crater.

(1) ⌞4.0/2.5;<0.2⌟ (2) ⌞1.3;<0.1⌟ (3) ⌞3.0;<0.2⌟ (4) ⌞2.0;<0.2⌟ (5) ⌞1.0;<0.1⌟ These are parts of the linearments around this crater.

(6) ⌞5̇/1.5;0.3e⌟ A ridge, which is east bank of a crack through the north-west wall.

(7) ⌞2.0;<0.2⌟ A faint mound having a minute pit to its north-west side.

Davy, A

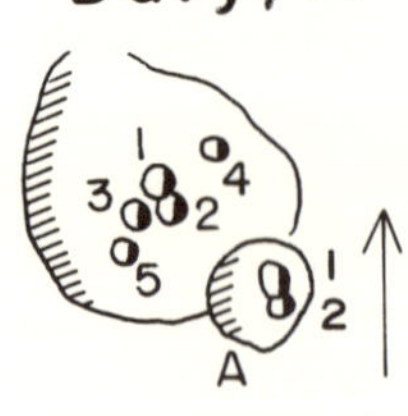

26. Davy (-138,-205;35) The floor is not flat.

(1) ⌞1.7;0.2⌟ (2) ⌞3/1.5;0.15⌟ (3) ⌞2.0̇;0.15⌟ These are triple mounds. The (1)

is likely to be the true central peak.

(4) [2.5;<0.1] A minute mound.

(5) [3.5;<0.2] Very low mound.

Delaunay

27. Davy A (-131,-212;15)

(1) [4.0;1.2w/0.3e] A large dome on the north-easterly eccentric position.

(2) [1.0;<0.1] A tiny dome on the south-east of the (1).

28. Delaunay (+041,-380;50)

(1) [29/11;1.5e/2.3w] A huge central ridge, dividing the floor into two parts, of which north-west side is lower.

(2) [4.0/1.5;<0.1] (3) [6.0/1.5;<0.1] (4) [3;<0.2]These are probably parts of a linear structure around this crater.

(5) [3;<0.1e] A round mound on the west bottom of the (1).

29. Donati (+084,-353;36)

(1) [13/7;1.3] A central peak having conical form, from which, to the north-west direction, a ridge runs.

30. Faye (+062,-365;37)

(1) [5.5;1.0] The central peak is somewhat oblong along north-west direction.

31. Gemma Frisius (+191,-564;88)

(1) [8;0.6] The central peak having conical form.

(2) [1.4;<0.1] (3) [1.0;U. L.] Tiny mounds.

(4) [5/2.5;<0.2] A ridge having uncinate form. It seems like a part of lava flow.

(5) [2.5;0.1] (6) [3;0.1] (7) [2;<0.1] These are likely to be remnant of collapsed wall.

Gemma Frisius

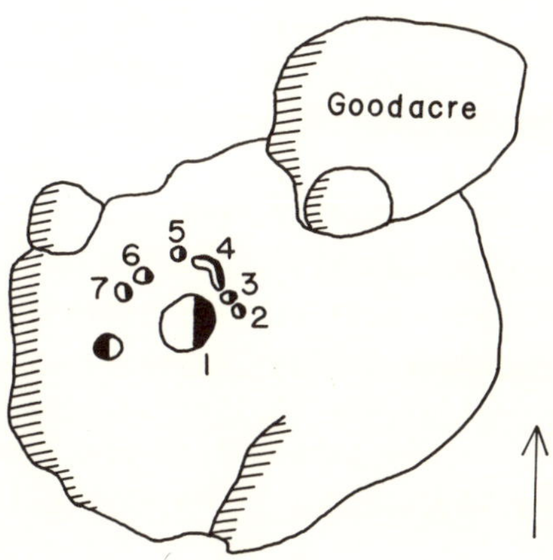

32. Gemma Frisius, U. F. on the W. Wall of (+161,-570;46)

(1) [11;<0.3] A broad central mountain, which is superposed by the south wall of Gemma Frisius D.

33. Gemma Frisius D (+156,-564;27)

(1) [5;0.3] A broad elevation.

34. Gemma Frisius EB, U. F. to the E. of (+195,-600;15)

(1) [3;U. L.] A slightly elevated part on the bottom of the south wall.

35. Gemma Frisius EB, U. F. to the N. of (+184,-594;17)

(1) [5.0/2.5;0.1] A central mountain, oblong in north-south line.

(2) [2.0;<0.1] A tiny mound to the west of the (1).

36. Gemma Frisius H (+179,-536;28)

(1) [5/2.5;0.15] (2) [2.5;0.1] The (1) is a mountain having uncinate form, and is likely to form a floor crater with the peak (2). Another viewpoint that the peak (2) is the true central peak, is also possible.

(3) [2.5;0.1] (4) [1.5;<0.1] Minute mounds.

U.F. to the North of Gemma Frisius EB

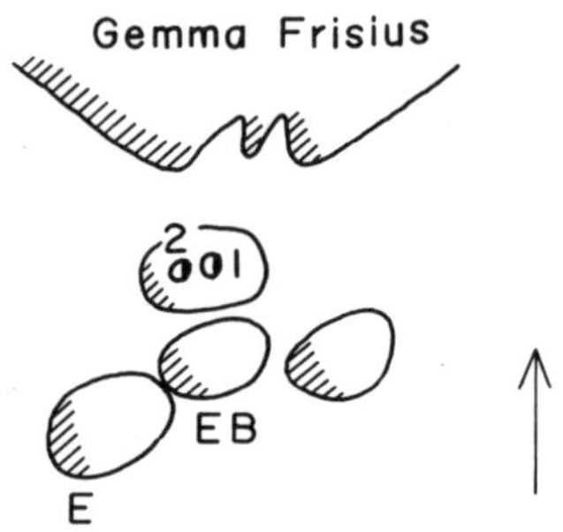

(VIII)

37. Gemma Frisius H, U. F. on the N. of (+177,-525;36) This formation is registered in the Arizona Catalogue as superposition of two craters (41572: ξ =+0.174, η =-0.525; 41582A: ξ =+0.181, η =-0.524). But it is likely to be a crater superposed by Gemma Frisius H, having a flat floor.

Gemma Frisius H

(1) ⌞1.4;0.5⌟ A tiny central hill.

38. Hell (-114,-535;33)

Hell

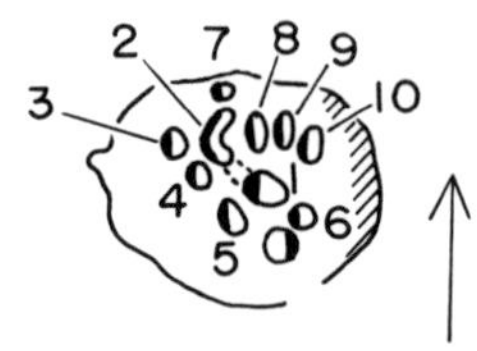

(1) ⌞5.5;1.0̇⌟ A peak on the easterly eccentric position of the floor. This is connected by a white ridge with the (2) and the (5).

(2) ⌞6/2;0.2⌟ A mountain having a form of uncinate ridge.

(3) ⌞2.8;0.6⌟ (4) ⌞3.5̇;0.5̇⌟ (5) ⌞2.7;0.7e⌟ (6) ⌞3.5̇;0.3⌟ Eminent mounds. Between the (5) and (6), there is a floor crater whose diameter is about 4 km.

(7) ⌞2.3;<0.2⌟ A very tiny mound on the slope of north wall.

(8) ⌞6/3;<0.2⌟ (9) ⌞6/3;<0.2⌟ (10) ⌞4/3;0.2⌟ Low ridges.

39. Hell A (-121,-557;22) The floor is rugged.

(1) ⌞3.5;0.4⌟ A typical central peak.

40. Hell B (-087,-500;22)

(1) ⌞1.0;0.1⌟ A very tiny mound on easterly eccentic position of the very smooth floor.

41. Hell J (-104,-495;6)

(1) ⌞2.5;U. L.⌟ A central pit whose diameter is about 2 km.

42. Klein (+044,-207;44)

(1) ⌞6/3;0.8⌟ Very oblong central mountain, whose direction is parallel to the linear structure around this crater.

43. La Caille E (+044,-398;27)

(1) ⌞6.0;0.9w/0.6e⌟ The central peak stands on the southerly eccentric position, having long slope towards west, and short slope towards east. From the peak, a long dyke runs to the south-easterly direction.

La Caille GA

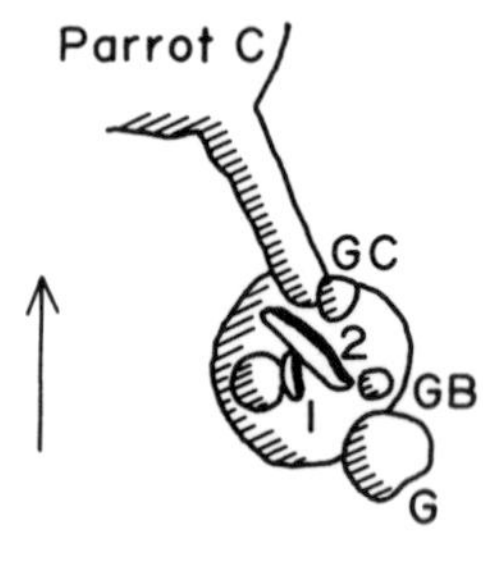

44. La Caille GA (+028,-344;24)

(1) ⌞4;0.7⌟ A mountain, superposed by another crater, whose diameter is 5 km. (This crater is registered as 40324A in the Arizona Catalogue.)

(2) ⌞12/4; U. L.⌟ A faint dyke runs on the floor from south-east to north-west.

45. Lexell (-059,-584;63) The floor is very rugged.

(1) ⌞7;0.6⌟ A central mountain, superposed by an oblong crater, whose shorter

diameter is 3.5 km.

(2) ⌞12/7;0.5̇⌟ A ridge, which is a part of collapsed wall.

Lexell

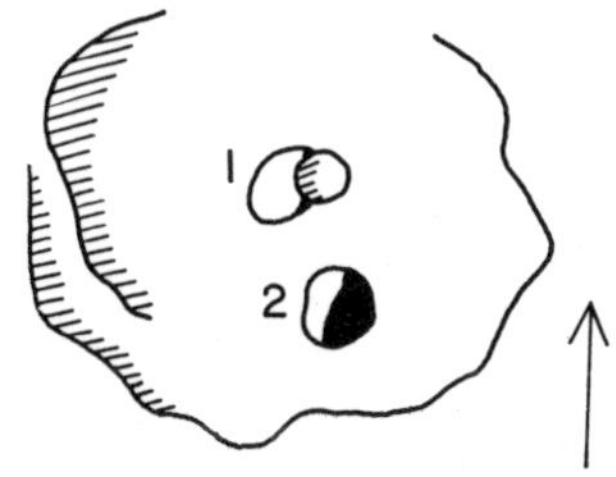

46. Lexell A (-018,-600;33)

(1) ⌞3.7;0.7⌟ A steep central peak.

(2) ⌞3.5;0.4⌟ A mound lower than the (1), situated to the east of it.

Lexell A

47. Nonius K (+058,-554;18)

(1) ⌞7/3;<0.2⌟ The ridge (1) is a part of the linear structure around this crater.

(2) ⌞2;U. L.⌟ A faint ridge which is also parallel to this structure.

48. Nonius L (+052,-552;31)

(1) ⌞8;0.2w⌟ A ridge, probably a part of collapsed wall.

(2) ⌞4;0.8w⌟ Probably the true central peak, superposed by the wall of Nonius K.

Nonius L, K

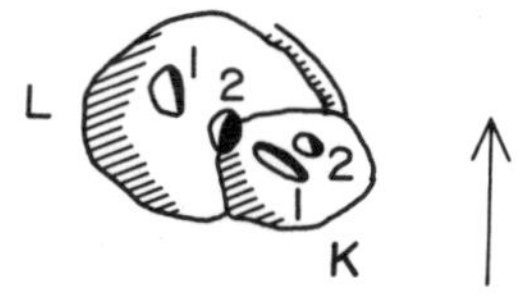

49. Parrot (+057,-252;68)

(1) ⌞2.7;0.6e⌟ The true central peak, superposed by the wall of Parrot D.

(2) ⌞2.0;0.4e⌟ (3) ⌞2.0;0.4w⌟ The (1), (2) and (3) are triple peak on a triangular plateau.

(4) ⌞4.5;0.2⌟ A distinct mound.

(5) ⌞10/2.5;0.2⌟ A ridge neighbouring the (4).

(6) ⌞8/6;0.1⌟ (7) ⌞8/6;0.1⌟ Broad plateaus.

(8) ⌞7/5̇;<0.1⌟ A bank along a long valley intruding through the wall of Parrot.

(9) ⌞10/5;<0.1⌟ A faint ridge parallel to the valley.

(10) ⌞7/4;<0.1⌟ A low hill.

Parrot, D

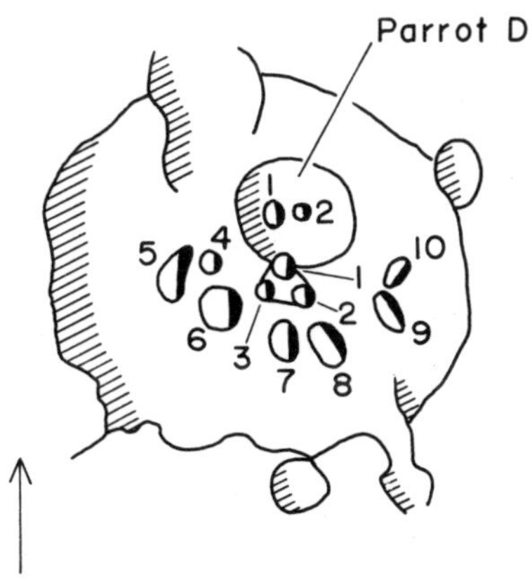

50. Parrot C (+020,-318;31)

(1) ⌞5.5/4.0;0.7⌟ An oblong central mountain.

(2) ⌞7/4;U. L.⌟ (3) ⌞5̇/3;U. L.⌟ Swelling parts of the floor.

Parrot C

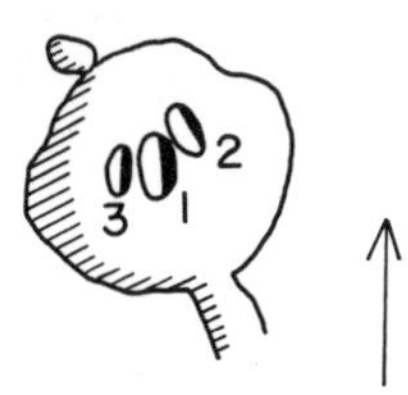

51. Parrot D (+061,-245;21)

(1) ⌞2.5;0.5̇⌟ A small peak which may be the true central peak.

(2) ⌞4;0.6⌟ A mountain. It also seems that these peaks form both sides of a valley.

52. Parrot E (+038,-275;20)

(1) ⌞4;U. L.⌟ Very faint, but broad mountain block.

53. Parrot F, U. F. Containing (+017,-278;41)

(1) ⌞4;0.6⌟ A central peak, superposed by the wall of Parrot F.

54. Parrot J (+030,-292;23) A valley runs through north-east part of the floor.

(1) ⌞5;<0.3⌟ Broad and low central mountain.

55. Pitatus G (-171,-497;16)

(1) ⌞1;0.2⌟ A very feeble central hill.

56. Playfair, U. F. to the N.-E. of (+150,-380;18)
On the south-east part of the floor, two lines of linear structure superpose on a large mountain block, whose diameter is about 8 km. On the north bottom of the block, there is a small shallow pit, whose diameter is about 3 km.
(1) ⌞1.5̇;0.2e⌟ A peak standing at the center of the floor. May be the true central peak.
(2) ⌞3;0.2e⌟ (3) ⌞3.5;0.2w⌟ Mounds standing on the linear structure.

U.F. to the N.E. of Playfair

57. Playfair G, U. F. on the N. Wall of (+104,-385;14)
(1) ⌞2.5;U. L.⌟ A small central patch is perceivable under special angle of illumination.

58. Poisson, U. F. Containing (+149,-509;84)
This formation is usually regarded as aggregate of several craters, such as Poisson and Poisson N, R, and T. Large dyke, whose width is 15 km at maximum, runs from north-east to south west.
(1) ⌞15;1.5⌟ The highest part of the dyke.
(2) ⌞6;0.9⌟ A peak having conical form. May be the true central peak.
(3) ⌞10;1.0⌟ Another higher peak on the dyke.

U.F. Containing Poisson

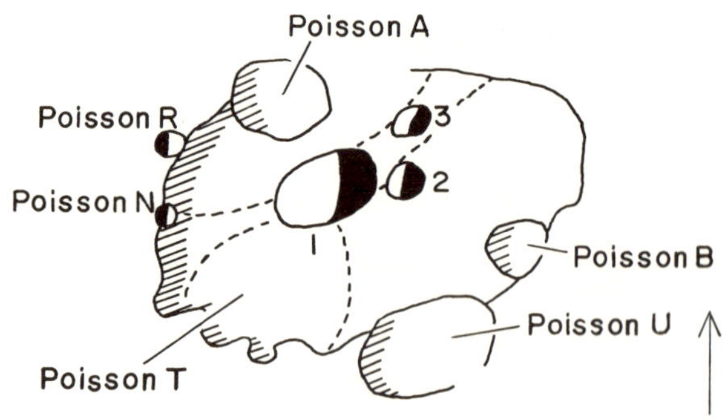

59. Poisson, U. F. to the E. of (+181,-513;46)
(1) ⌞4;0.6⌟ Very tiny central hill, which looks like a white patch under high illumination.

60. Poisson J (+120,-574;29)
(1) ⌞1.5;0.2e/0.5̇w⌟ On northernly eccentric position of the floor, there is a hill, which is somewhat oblong to north-south direction. It forms east bank of a small pit whose diameter is about 2 km.

61. Poisson U (+152,-524;25)
This crater is superposed by Poisson V. Through center of it, a long, uncinate ridge, whose length is about 15 km, is seen.
(1) ⌞6/4;0.2e/1.0w⌟ (2) ⌞5̇/4;0.2e/1.2w⌟ (3) ⌞3.5;0.4e/1.0w⌟ On the ridge above mentioned, stand three peaks, having wider slope to the west. Among them, the peak (3) is most distinct.

Poisson U

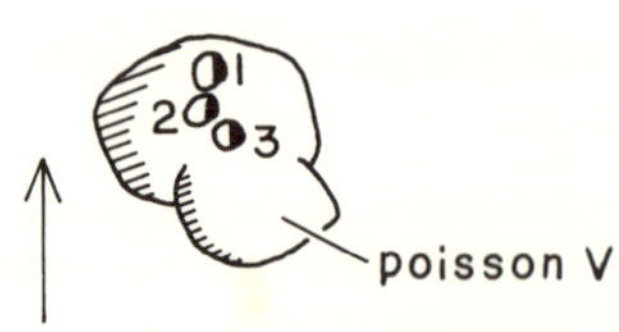

62. Purbach (-030,-430;118)
(1) ⌞6.5;0.7⌟ (2) ⌞7.0;0.7⌟ (3) ⌞1.5;0.2⌟ (4) ⌞5̇;0.3⌟
(5) ⌞12/5̇;0.5⌟ (6) ⌞7/5̇;0.6⌟ (7) ⌞2̇0/5̇;0.6⌟
The ridges (1), (2), (3), (4), (6) and south

part of (5) form the wall of collapsed crater, Purbach W. The ridges (5) and (7) are likely to be the wall of another collapsed crater. Between the (5) and (7), there is a rugged part of floor. To the south-west of the floor crater, Purbach A, there is another rugged part of floor.

(8) [2;0.1] Very tiny peak. But it looks like the true central peak.

(9) [2.5;U. L.] (10) [3;U. L.] Very faint and broad elevations.

Purbach

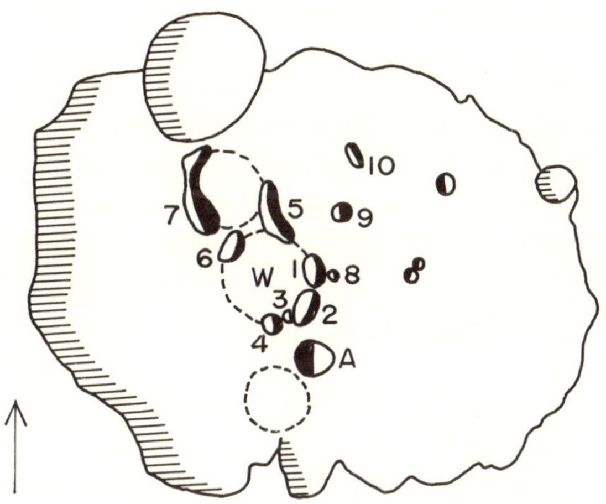

63. Purbach, U. F. on the N.-W. Wall of (-064,-410;8)
    (1) [2.0;<0.1] Very faint central mound.

64. Purbach DB (-023,-376;19)
    (1) [7.5/3.5;0.25e/0.35w] Very long ridge, which looks very whitish under high illumination, runs from north-east to south-west. The highest part of this ridge is westerly deviated.

65. Purbach FA (+000,-411;14)
    (1) [4;U. L.] There is a broad elevation on the southerly eccentric position of the floor.

66. Regiomontanus (-016,-477;126/110)
    (1) [15;1.2] A very large central hill, having a summit crater whose diameter is 5.5 km.
    (2) [7;0.5] (3) [3;0.2] Attached hills of the (1).
    (4) [1.7;0.1] A tiny mound.

67. Regiomontanus, U. F. to the S. of (-011,-503;12)
    (1) [7/5;0.2] A central hill slightly oblong to north-south direction.

Regiomontanus

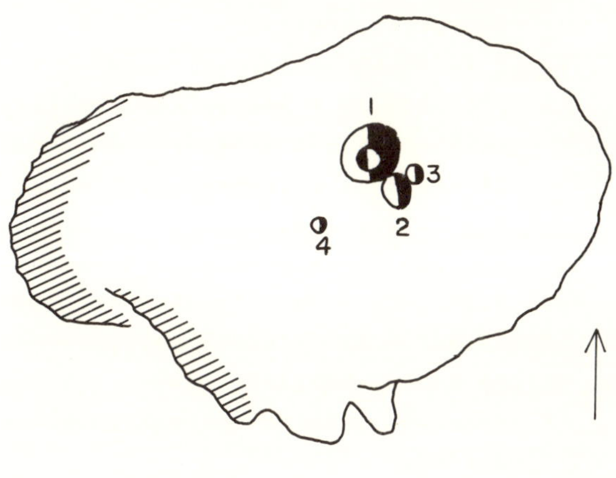

68. Ritchey,U. F. to the S.-E. of (+167,-205;45/25)
    This formation is ordinarily registered as a compound of four craters containing Ritchey J.
    (1) [4/1.5;<0.1] Very faint ridge.
    (2) [5.5/2.5;0.15] (3) [5.5/2.5;0.13] Ridges parallel to each other.

(4) ⌞2;0.1⌟ A tiny mound.
(5) ⌞4;0.25⌟ A distinct peak.
(6) ⌞8/2.5;<0.15⌟ (7) ⌞8/2.5;0.15⌟ Uncinate ridges.
(8) ⌞1.5;<0.1⌟ (9) ⌞1;<0.1⌟ Tiny hills.
(10) ⌞2;<0.1⌟ A faint mound.
(11) ⌞2.8;<0.1⌟ A round mound.
(12) ⌞1.7;<0.1⌟ Another round mound.
(13) ⌞5;<0.1⌟ A broad elevation.
(14) ⌞4.5/2.5;<0.1⌟ Somewhat oblong hill.

U.F. to the S.E. of Ritchey

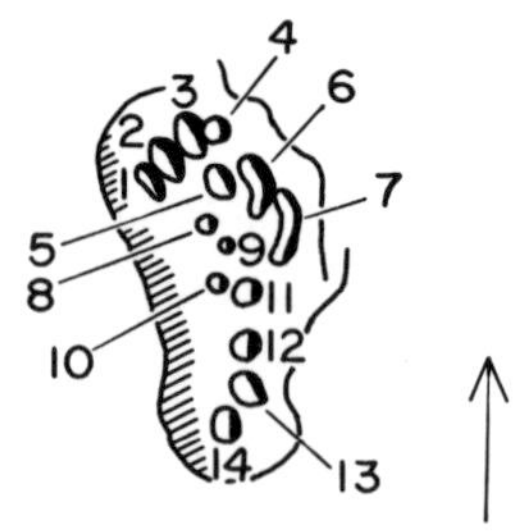

69. Thebit (-064, -375;55)
(1): x ⌞5.5;0.5́⌟ A peak standing on the bottom of north-east wall. From it, an uncinate groove runs to west.
(2) ⌞1.5;0.1⌟ (3) ⌞2.5;0.2⌟ (4) ⌞1;<0.1⌟ Along the above mentioned groove, stand three mounds (2), (3) and (4), among which the most distinct one is the (3).
(5) ⌞2;<0.1⌟ A white ridge.
(6) ⌞3.5;<0.1⌟ A dark elevation.
(7) ⌞4/2;0.1⌟ (8) ⌞5/2.5;0.25⌟ (9) ⌞4/2.5;0.1e⌟ Ridges, almost parallel to each other.

Thebit, A, L

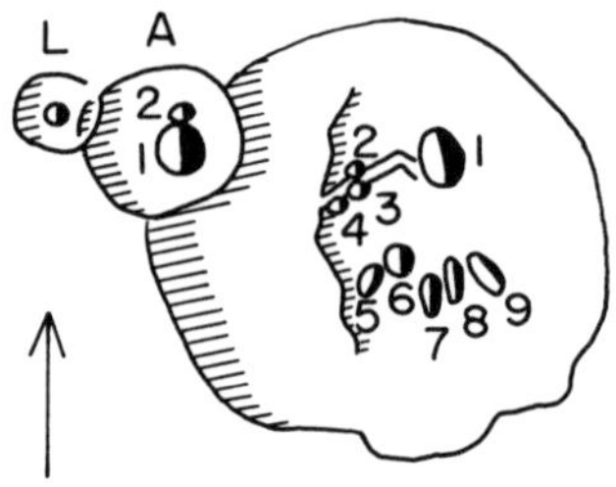

70. Thebit A (-079,-368;20)
(1) ⌞5̇;<0.5⌟ Centrally elevated floor.
(2) ⌞2.5;<0.5⌟ A tiny round mound on the bottom of the north-wall.

71. Thebit L (-087,-366;10) Its south-east wall superposes on the west wall of Thebit A.
(1) ⌞2.5;0.2⌟ A tiny mound on the somewhat southerly eccentric position.

Vogel, B

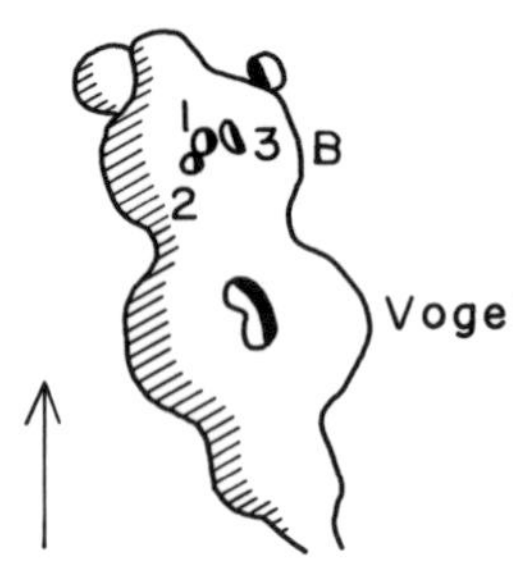

72. Vogel (+099,-260;27)
(1) ⌞8/2.5;0.5̇⌟ A gentle and uncinate hill on the north-easterly eccentric position of the floor. It seems as a remnant wall of an old crater.

73. Vogel B (+096,-248;22)
(1) ⌞3;0.3⌟ The most distinct peak.
(2) ⌞2;U. L.⌟ A very faint peak, but it seems as the true central peak.
(3) ⌞3.5/2;0.2⌟ A faint fault.

74. Walter (+010,-545;140/132)
(1) ⌞11;2.0⌟ A large central mountain having a summit crater whose diameter is 2.2 km.
(2) ⌞11/7;0.8⌟ A broad hill.
(3) ⌞4;0.6⌟ A conical peak.

(4) ⌞5̇;U. L.⌟ A low hill.

(5) ⌞2;0.1⌟ (6) ⌞2;0.3⌟ (7) ⌞4/2;0.2⌟ Hills which form a wall of a collapsed crater.

(8) ⌞9;0.4⌟ A mountain having a summit crater whose diameter is about 4 km.

(9) ⌞3;le⌟ (10) ⌞3;0.3⌟ Elevated parts of rugged floor.

(11) ⌞1.6;0.2⌟ A round mound.

(12) ⌞14/3;0.5⌟ (13) ⌞5̇/4;0.3⌟ Ridges on the rugged part of floor.

(14) ⌞3;0.3⌟ A tiny peak.

(15) ⌞3.5/2.5;0.3⌟ An oblong hill which is an extension of the ridge (13).

Walter

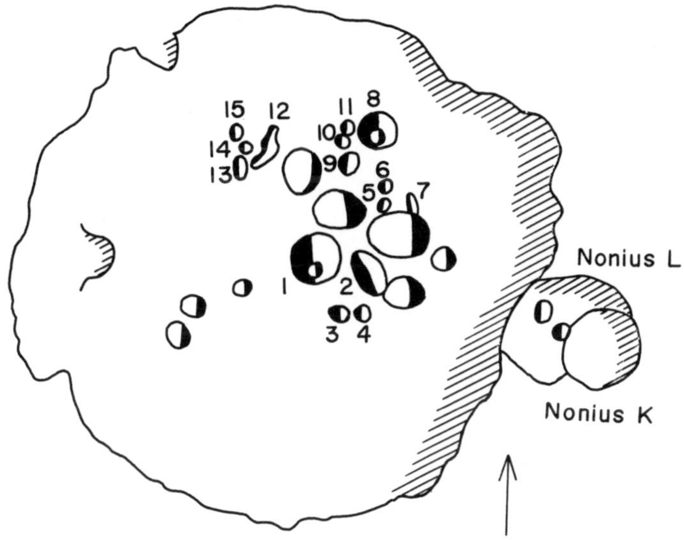

75. Walter, U. F. on the Floor of (+024,-542;14)

(1) ⌞1.5̇;U. L.⌟ A faint white elevation on the floor.

76. Werner (+051,-470;70)

(1) ⌞2.3;0.4⌟ (2) ⌞3;0.8⌟ Higher parts of a ridge.

(3) ⌞5̇/3.5;0.2⌟ An oblong peak. May be the true central crater.

(4) ⌞3.9;0.8⌟ The highest peak.

(5) ⌞2.5;0.4⌟ Very faint peak.

77. Werner, U. F. to the E. of (+084,-462;31)

(1) ⌞4;<0.8⌟ This formation is very indistinctive. But the central mound is rather remarkable.

78. Werner H (+024,-449;16)

(1) ⌞3;U. L.⌟ Very faintly centrally elevated floor.

Werner

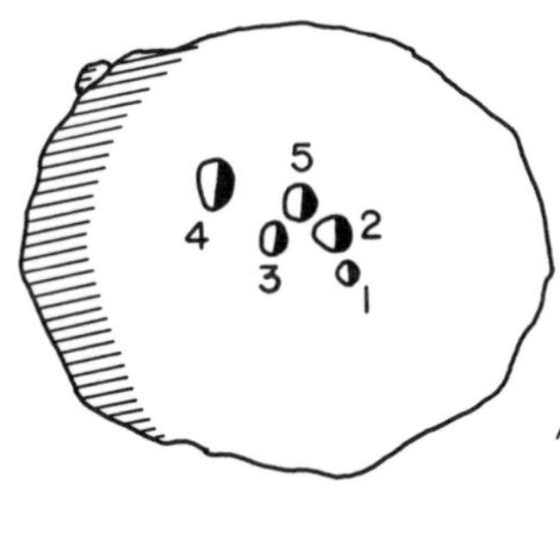

IX

Abulfeda

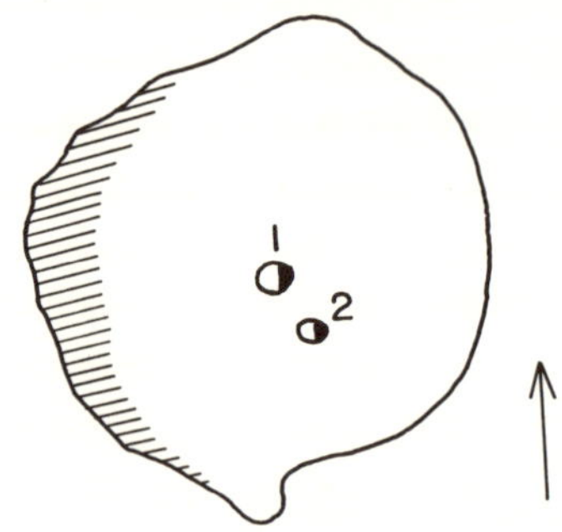

U.F. to the East of Abulfeda BA

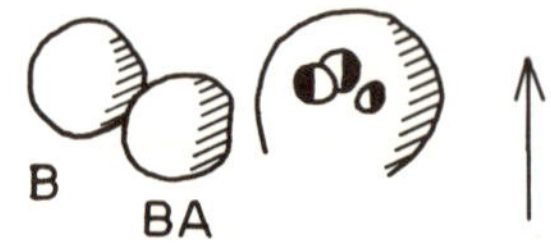

Beaumont

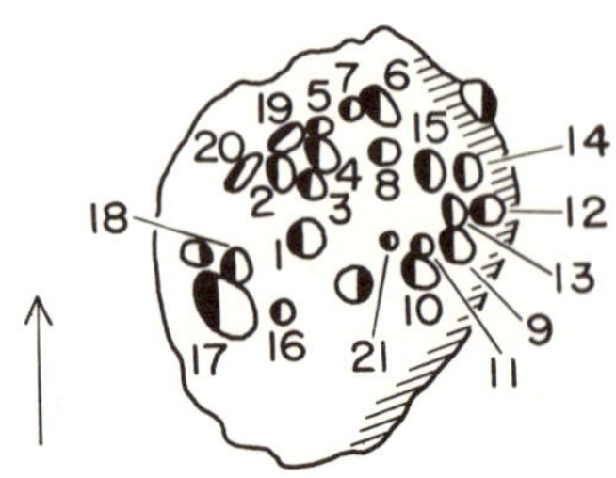

1. Abulfeda (+234,-239;62)
   (1) ⌞3.8;U.L.⌟ A white patch. Probably remnant of the central peak.
   (2) ⌞2.9;0.1⌟ A square-formed plateau.
2. Abulfeda BA, U. F. to the E. of (+293,-253;21)
   This formation has an elevated floor like that of Wargentin.
   (1) ⌞3;U. L.⌟ This peak forms the west rim of a small pit whose diameter is about 3.5 km.
3. Almanon B (+250,-314;25)
   (1) ⌞2.7;1.1⌟ Very distinct central peak.
4. Almanon P, U. F. Containing (+278,-318;19)
   (1) ⌞4;1w⌟ This peak is the highest part of the rampart of Almanon P.
5. Beaumont (+458,-310;53) The floor is very rugged.
   (1) ⌞4;0.3⌟ A peak having a conical form.
   (2) ⌞4/2;0.3⌟ (3) ⌞4;0.2⌟ (4) ⌞4/3.5;0.5̇w⌟ (5) ⌞2;0.1⌟ (6) ⌞5̇/4;0.3⌟ (7) ⌞1.5;0.1⌟ (8) ⌞3;0.2⌟ (9)⌞5̇/3; 0.2⌟ (10) ⌞5̇/3;<0.2⌟ (11) ⌞2;0.1⌟ (12) ⌞2.5;0.3⌟
   (13) ⌞7/2.5;<0.2⌟ The oblong mountains (2), (3), (4), (6), (9), (10) and (12) are all parallel to the linear structure around this crater. The mountain (4) is the highest. The (5) and (7) are attached hills of the (4) and (7) respectively. The (8) is a rather distinct hill. The (11) is a faint elevation between the (10) and (9). The (13) is an oblong faint elevation, but is not parallel to the (9) and (12).
   (14) ⌞5̇/3;<0.2⌟ (15) ⌞5̇/3;<0.2⌟ Very faint elevations.
   (16) ⌞1.5;0.1⌟ A tiny mound.
   (17) ⌞10/7;0.3⌟ A very broad elevation.
   (18) ⌞2.7; 0.2⌟ A faint ridge.
   (19) ⌞4.5/3.5;0.2w⌟ (20) ⌞5.5/2.5;<0.2⌟ Oblong hills parallel to the north-west wall of this crater.
   (21) ⌞ 2;0.1⌟ A very tiny mound.
6. Beaumont, U. F. to the S.-E. of (+471,-330;12)
   (1) ⌞5;<0.3⌟ Very broad central mountain, which looks like a part of the wall of superposed crater.
7. Beaumont A (+447,-280;12)
   (1) ⌞5.5/4;0.4w/0.1e⌟ Very large, oblong mountain on the easterly eccentric position of the floor.
8. Beaumont B, U. F. on the W. of (+417,-319;15)
   (1) ⌞4;<0.2⌟ Large and low central mountain.

(IX)

9. Catharina (+381,-311;97)

(1) ⌞7;0.8⌟ The central peak is superposed by the wall of Catharina P.

10. Catharina B (+394,-292;25/22)

(1) ⌞5̇/4;0.2⌟ Somewhat oblong ridge on the westerly eccentric position of the floor.

11. Catharina C (+387,-348;30)

The east and west border of the floor are depressed. Therefore the floor has a plateau-like form. Its surface is rugged.

(1) ⌞3;0.2⌟ A small peak on its center.

12. Catharina P (+378,-294;49)

(1) ⌞3.5;0.4⌟ Very distinct peak near the south wall of Catharina P.

(2) ⌞2.5;0.3⌟ A peak on the easterly eccentric position of the floor.

13. Celsius, U. F. to the W. of (+257,-551;17)

(1) ⌞5̇/4;0.8w/0.5e⌟ (2) ⌞6/4;0.5⌟ Two oblong hills lying on the north and south slope of the wall respectively. The (1) is more distinct.

Catharina, B, P

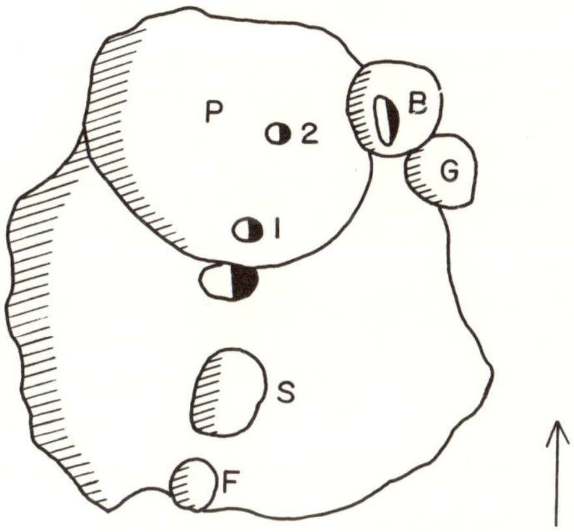

U.F. to the West of Celsius

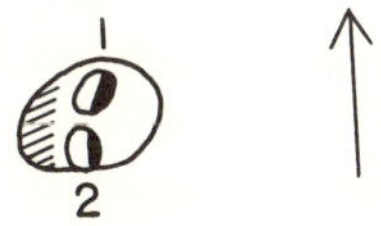

Cyrillus

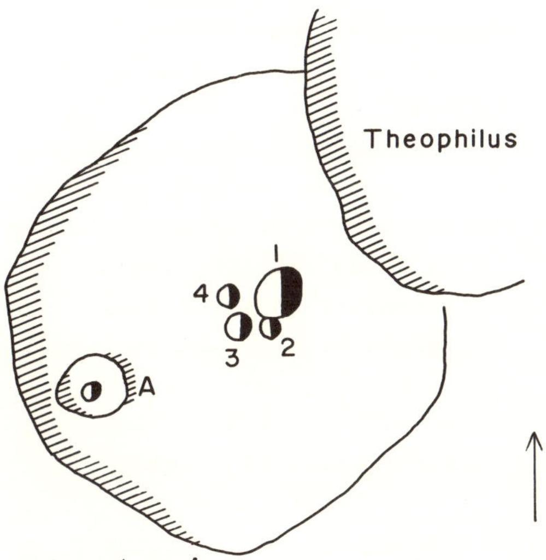

14. Cyrillus (+397, -231; 93)

(1):α ⌞10̇;1.6⌟ The largest peak, situated on the north-westerly eccentric position, having conical form.

(2) ⌞4.5;0.5̇⌟ Small attached peak.

(3):δ ⌞8;1.2⌟ May be the true central peak.

(4):η ⌞5.5;0.6⌟ A mound.

15. Cyrillus, U. F. to the S.-W. of (+355,-258;20)

(1) ⌞7/4;0.5̇⌟ A gentle ridge on the south-westerly eccentric position of the floor.

16. Cyrillus A (+381,-238;15)

(1) ⌞3.5;U. L.⌟ Centrally elevated floor.

17. Cyrillus F, U. F. on the E. Wall of (+426,-266;22)

(1) ⌞8;0.6⌟ A large, but very low mound, whose summit is faintly depressed.

18. Daguerre (+541,-205;45)

(1) ⌞5;<0.1⌟ Very broad and low mound.

(2) ⌞2.5;<0.1⌟ (3) ⌞2.5;<0.1⌟ Hills which are higher part of wall of central

(IX)

crater, whose diameter is about 4 km.

Fermat

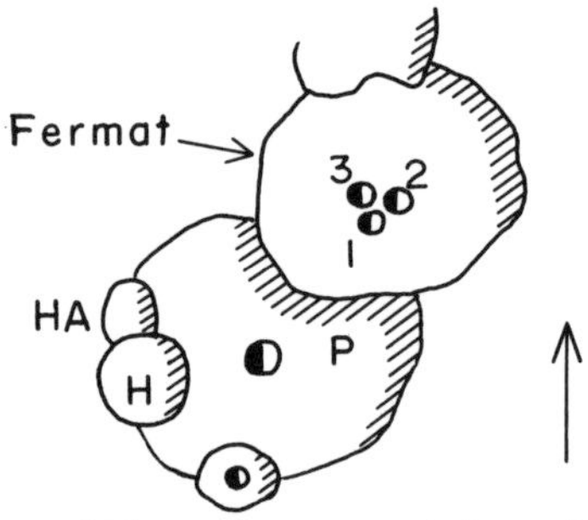

19. Fermat (+313,-385;39)
On the east part of the floor, a faint crack runs.
(1) ⌜3;U. L.⌟ (2) ⌜3;U. L.⌟ (3) ⌜3;U. L.⌟ On the south part of the floor, three white patches are perceived, which Wilkins and Moore described as hills.

Daguerre

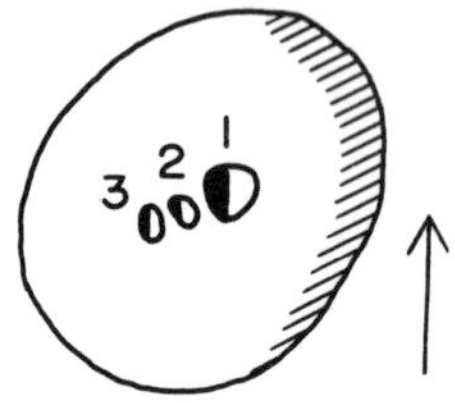

20. Fermat C, U. F. on the E. of (+305,-361;31)
(1) ⌜7/4;0.4⌟ (2) ⌜7/4.5;0.3⌟ A ridge runs through the floor. Its highest part is the (1), and the (2) is connected with it.

21. Fermat P (+304,-400;36)
(1) ⌜4.4;0.5̇⌟ A round mound.

22. Fermat P,U. F. on the S. Wall of (+301,-410;8)
(1) ⌜3;<0.1⌟ Square-formed crater, having a central mound.

U.F. on the East of Fermat C

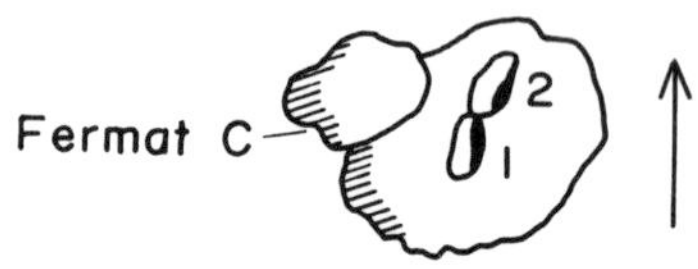

Fracastorius,D

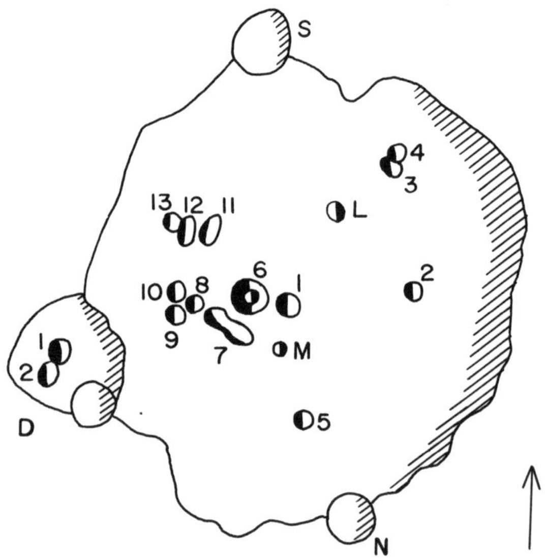

23. Fracastorius (+509,-363;124)
(1) ⌜4;0.6⌟ The most distinct conical peak.
(2) ⌜4;0.3⌟ Isolated peak.
(3) ⌜4;0.3⌟ (4) ⌜3;0.2⌟ Twin mounds.
(5) ⌜3.5;0.3⌟ Isolated peak.
(6) ⌜8;0.2⌟ A broad mound having a summit crater, whose diameter is about 4 km.
(7) ⌜10/4;U. L.⌟ A faint ridge.
(8) ⌜3;U. L.⌟ (9) ⌜5̇;U. L.⌟
(10) ⌜4;U. L.⌟ (11) ⌜7/5̇;U. L.⌟
(12) ⌜5̇;U. L.⌟ (13) ⌜6;U. L.⌟
Very faint elevations.

24. Fracastorius, U. F. on the N. of (+505,-329;12)
(1) ⌜4;U. L.⌟ Centrally elevated floor.

25. Fracastorius A (+542,-413;18)
(1) ⌜5̇;U. L.⌟ The floor is very faintly elevated.

26. Fracastorius A, U. F. to the N.-W. of (+533,-407;16)
(1) ⌜4;U. L.⌟ A faint ridge.

27. Fracastorius B (+560,-384;27)
(1) ⌜6.5;0.9⌟ Large conical central peak.

(IX)

28. Fracastorius D (+477,-371;27)

(1) ⌞5;0.7⌝ (2) ⌞5.5;0.5̇⌝ Superposing on the central peak (1), a ridge runs through the floor, of which main part is the (2). This is an extension of the ridge running to the south-west exterior of this crater.

29. Fracastorius K, U. F. on the S. of

(+521,-436;27) This formation is consist of superpositions of many craters.

(1) ⌞4;0.4⌝ A round mound.

(2) ⌞10;0.7⌝ A long ridge.

U.F. on the South of Fracastorius K

30. Gaudibert H (+581,-238;11)

(1) ⌞3;U. L.⌝ Very low mound.

31. Gemma Frisius, U. F. on the N.-E. Floor of

(+209,-556;12)

(1) ⌞3;<0.1⌝ Very low central mound.

32. Goodacre (+205,-540;46)

(1) ⌞4;0.6⌝ A typical central peak.

33. Goodacre G (+201,-549;16)

(1) ⌞4;<0.2⌝ A low central mound.

34. Goodacre K (+200,-514;11)

(1) ⌞2;<0.1⌝ Very minute central peak.

35. Kant Q, U. F. on the S.-E. of (+316,-228;9)

(1) ⌞6/3;U. L.⌝ A long linearment penetrates this formation, and superposes with the central peak.

36. Lindenau (+356,-534;53)

(1) ⌞5;0.6⌝ A conical peak.

(2) ⌞9.5;0.5̇⌝ A broad mountain.

(3) ⌞6;0.1⌝ A low hill.

(4) ⌞7/4;0.4e⌝ (5) ⌞5/3;<0.1e⌝ The (4) is a low ridge, with which another ridge (5) is connected.

Lindenau

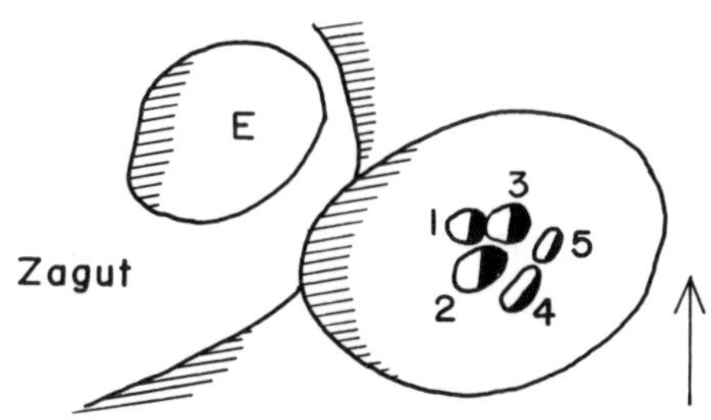

37. Lindenau H (+379,-520;11)

(1) ⌞1;U. L.⌝ Very faint and minute elevation on the southerly eccentric position of the floor.

38. Neander (+549,-520;52)

(1) ⌞7;1.3⌝ Large central peak.

(2) ⌞5̇;1.1⌝ Another peak on the west of the (1).

(3) ⌞5̇;0.7⌝ (4) ⌞3;0.4⌝ Small mounds.

(5) ⌞6;<0.4⌝ A broad elevation on the east part of the floor.

Neander

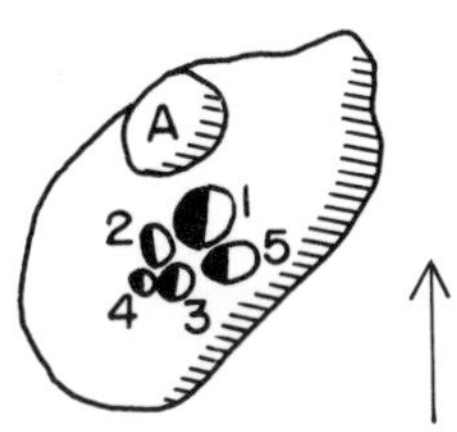

39. Neander C (+516,-479;20)

Probably this formation is a compound crater.

(1) ⌞4;U. L.⌝ Very faint elevation on the southerly eccentric position of the floor.

40. Piccolomini (+464,-497;89)

(IX)

(1): δ⌞20;2.7w⌟Very lofty central mountain.

(2) ⌞8;1.1e⌟ (3) ⌞7;0.9⌟ On the east bottom of the (1), stand two hills (2) and (3).

(4) ⌞3;U. L.⌟ (5) ⌞3;U. L.⌟ Faint mounds on the north-east part of the floor.

(6) ⌞4;U. L.⌟ (7) ⌞4;U. L.⌟ Broad mounds to the east of the central mountain.

(8) ⌞5;U. L.⌟ A broad mound on the south of the central mountain.

(9) ⌞4;0.2⌟ Rather distinct hill on the bottom of the south-west wall.

(10) ⌞4;0.1⌟ (11) ⌞4;U. L.⌟ Very faint elevations.

(12) ⌞3;0.1⌟ (13) ⌞3;U. L.⌟ (14) ⌞2;0.1⌟

(15) ⌞3;U. L.⌟ Small mounds to the north-west of the central mountain.

Piccolomini

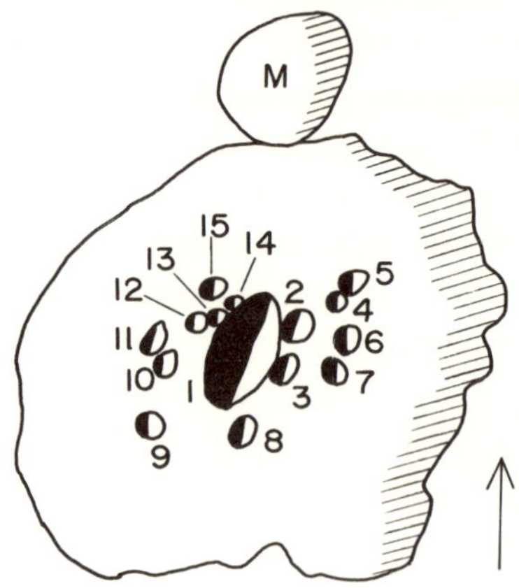

41. Piccolomini C (+459,-465;26)

(1) ⌞5;U. L.⌟ A broad elevation on the east wall.

(2) ⌞5;U. L.⌟ An elevated part of the floor.

(3) ⌞4;U. L.⌟ (4) ⌞3;U. L.⌟ Mounds on the north wall.

42. Piccolomini C, U. F. Superposed by (+453,-460;29)

(1) ⌞6;1.3w⌟ The central mountain, superposed by the west wall of Piccolomini C.

(2) ⌞4;0.6w⌟ Neighbouring mountain to the north of the (1).

Piccolomini C, M and U.F.

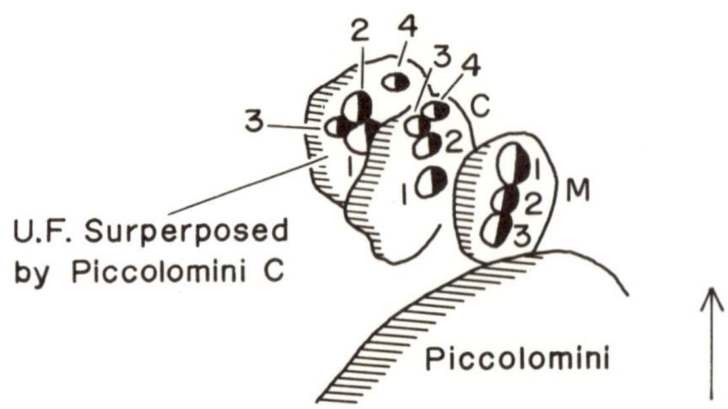

(3) ⌞2.5;0.1⌟ A tiny mound on the west wall of this formation.

(4) ⌞4;0.7w⌟ Broad hill on the north wall of this formation.

43. Piccolomini M (+467,-468;25)

(1) ⌞6;U. L.⌟ (2) ⌞5;U. L.⌟ (3) ⌞6;U. L.⌟ A broad ridge runs through this formation. The (1), (2) and (3) are higher parts of this ridge.

44. Piccolomini S (+479,-523;22)

(1) ⌞5;U. L.⌟ A low mound.

45. Polybius A,U. F. to the S. of (+434,-404;8)

This formation is not registered in the Arizona Catalogue.

(1) ⌞3;U. L.⌟ The central mountain is very minute.

46. Pons (+332,-428;41/31)

(1) ⌞15/8;2⌟ A long ridge, whose north part is highest.

(2) ⌞3;U. L.⌟ A faint elevation.

(3) ⌞10/6;<0.3⌟ A broad hill having a small crater whose diameter is about

3 km.

(4) ⌞6/4;0.5w⌟ A hill having westerly steep slope.

47. Pons C (+338,-469;18)

(1) ⌞3;U. L.⌟ A tiny central mound.

Pons

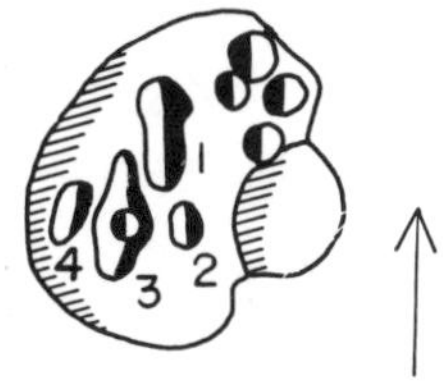

48. Pontanus (+220,-426;54)

The south part of the floor is very rich in parallel linearments.

(1) ⌞10;0.8⌟ A large central mountain whose north-west bottom is rugged.

(2) ⌞4;0.4⌟ A conical peak.

(3) ⌞13/3;U. L.⌟ A long low dyke.

(4) ⌞2;0.1⌟ A dark low hill.

(5) ⌞5̇;0.1⌟ A low peak.

(6) ⌞1.5;<0.1⌟ (7) ⌞1;<0.1⌟ Very minute white mounds.

Pontanus

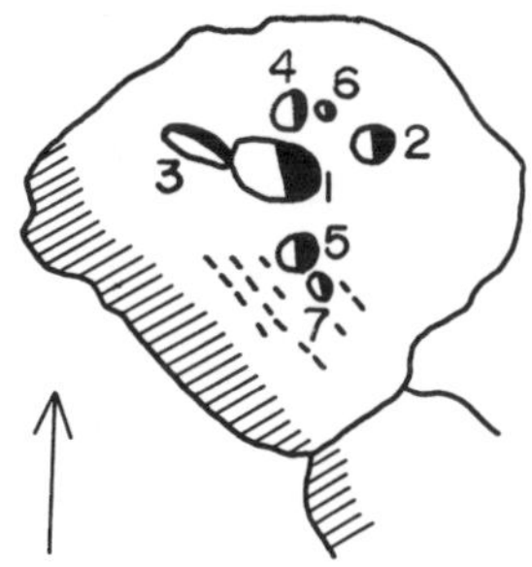

Pontanus C, CB, G

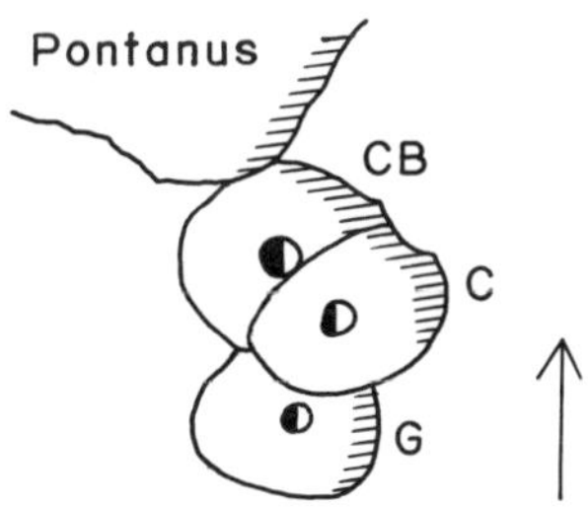

49. Pontanus C (+2̇31, -502;25)

(1) ⌞4;1.3⌟ A conical central peak.

50. Pontanus CB (+228, -497;26)

(1) ⌞4;0.8⌟ A central peak, superposed by the north-west wall of Pontanus C.

51. Pontanus D (+206,-437;20)

(1) ⌞5.5/4;<0.7⌟ A faint central elevation, which is oblong in north-south direction.

(2) ⌞3;U. L.⌟ A small mound on the north wall.

52. Pontanus E (+208,-426;13)

(1) ⌞10.0;U. L.⌟ A typical double crater.

53. Pontanus G (+227,-510;19)

(1) ⌞4;1.1w/0.6e⌟ A central mound halfly superposed by the south wall of Pontanus C.

Pontanus D, E

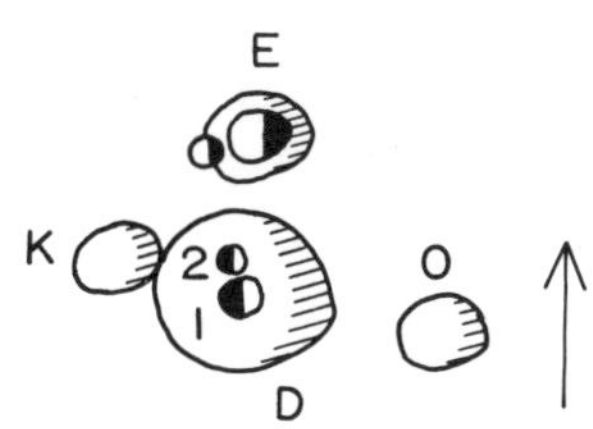

54. Pontanus H (+237,-521;30)

(1) ⌞5̇;0.7⌟ A typical central peak.

55. Rabbi Levi, U. F. to the W. of (+304,-570;46)

(1) ⌞1.5;U. L.⌟ Very minute peak on the east rim of small crater whose diameter is about 4 km.

56. Rheita, U. F. to the N.-W. of (+579,-590;14)

(1) ⌞3.5;U. L.⌟ Very low round elevation.

57. Rheita C, U. F. to the S.-E. of (+577,-581;35)

(1) ⌞5̇;0.9⌟ A round central mound.

58. Rothmann (+401,-513;43)

(1) ⌞10̇/4;0.3⌟ A long low ridge runs on the west part of the floor.

59. Rothmann B (+405,-527;18)

Sacrobosco

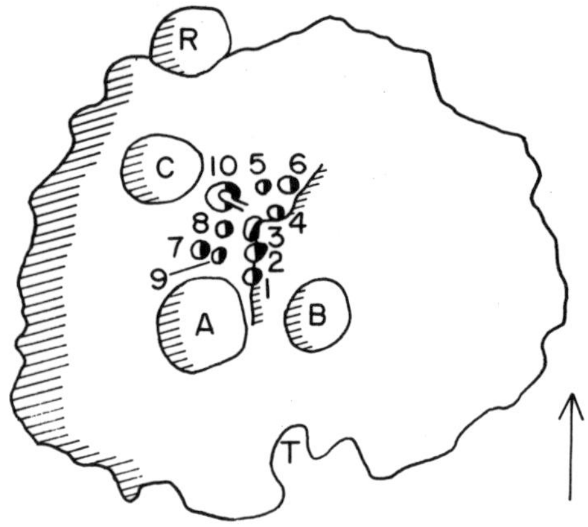

(IX)

(1) ⌞2.5;U. L.⌟ Very minute peak, probably the true central peak.

(2) ⌞5;U. L.⌟ Very faint elevation on the east part of the floor.

U.F. on the West of Rabbi Levi

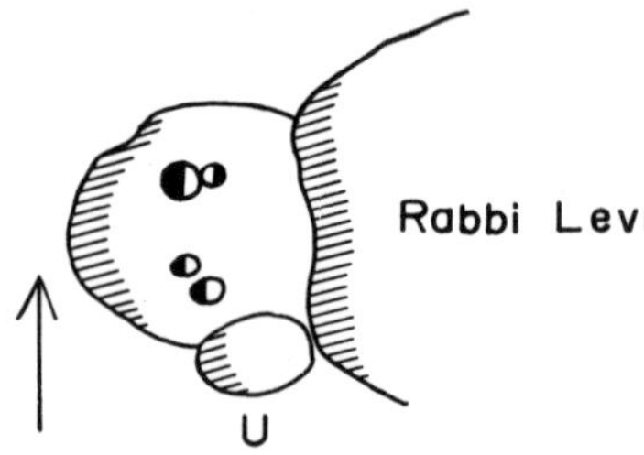

60. Rothmann C (+372,-479;20)
   (1) ⌞7;U. L.⌟ Very faintly elevated floor.

Rothmann B

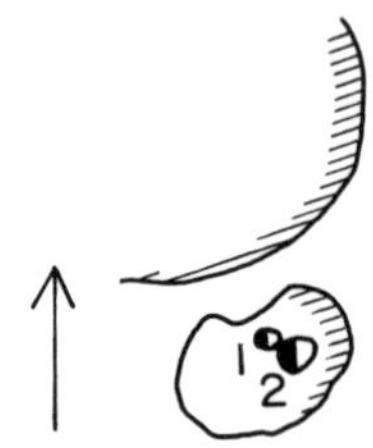

61. Sacrobosco (+263,-401;96)
   (1) ⌞2.5;0.4e⌟ A small peak which forms east rim of a small crater whose diameter is about 3.0 km.
   (2) ⌞3.5;0.5e⌟ Probably true central peak.
   (3) ⌞4/3;0.4e⌟ An oblong hill.
   (4) ⌞2.5;0.2e⌟ (5) ⌞3;0.2⌟ (6) ⌞3;0.2⌟ Small mound.
   (7) ⌞3;0.2⌟,A small mountain.
   (8) ⌞2.5;0.25⌟ A conical peak.
   (9) ⌞2;U. L.⌟ A small faint elevation.
   (10) ⌞5;U. L.⌟ A faint elevation having a summit crater, from which a groove runs towards south-east direction.
   A scarp runs on the floor in north-south direction, and the peak (1), (2), (3) and (4) stand on its edge.
62. Sacrobasco F (+270,-361;18)
   (1) ⌞5/4;U. L.⌟ Very low dyke runs from the north wall to the center of the floor.
63. Stiborius (+439,-565;43)
   (1) ⌞7;1.0⌟ A low conical central mountain.

Stiborius A

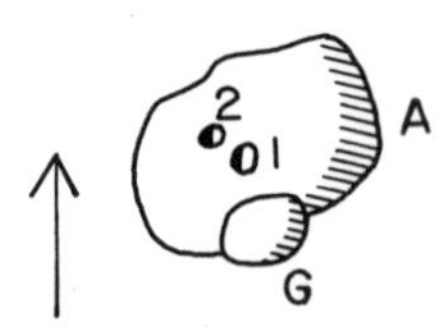

64. Stiborius A (+466,-600;32)
   (1) ⌞3;<0.3⌟ A faint mound on the south part of the floor.
   (2) ⌞2;U. L.⌟ A small and faint mound at the bottom of the west wall.
65. Stiborius C (+456,-558;22)
   (1) ⌞5;U. L.⌟ Very faintly centrally elevated floor.
66. Stiborius EA, U. F. to the E. of (+479,-573;15)
   (1) ⌞3.5/2;U. L.⌟ A low and tiny central mountain somewhat oblong along north-south direction.
67. Stiborius K (+462,-581;16)
   This formation is likely to be a compound crater.

(1) ⌜6/4;U. L.⌟ Very faint ridge on the south-east wall.

68. Stiborius K, U. F. to the N. of (+466,-574;10)

(1) ⌜1.0;U. L.⌟ A very faint and minute central mound.

69. Tacitus (+313,-279;40)

(1) ⌜13/3;0.5̇⌟ Very long dyke parallel to the meridian line.

(2) ⌜4;0.5̇⌟ A round mound.

(3) ⌜1.5;0.2⌟ A very minute mound.

(4) ⌜6/3;0.2e⌟ A hill, which is collapsed part of the east wall.

(5) ⌜4;U. L.⌟ (6) ⌜3.5;U. L.⌟ (7) ⌜4;U. L.⌟ Very faint elevations.

Tacitus

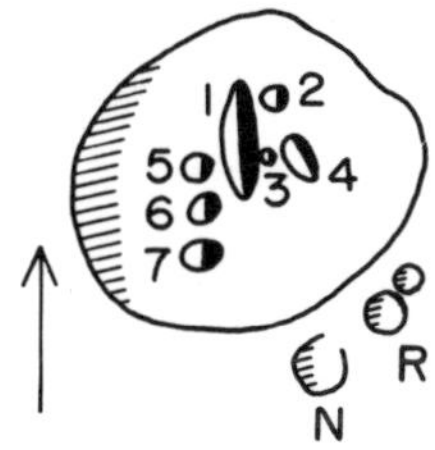

70. Tacitus D (+348,-234;21)

(1) ⌜5.5;0.5̇⌟ A mountain having a summit crater whose diameter is 1.6 km.

(2) ⌜7/3.5;0.5̇⌟ Collapsed part of the north wall.

(3) ⌜3/2.5;0.2⌟ A faint oblong hill.

Tacitus D

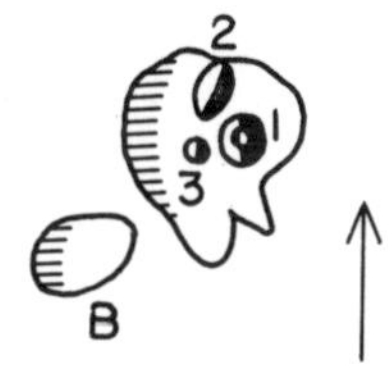

71. Weinek (+535,-462;32)

(1) ⌜4;0.2⌟ Very low central mound.

72. Zagut B (+272,-532;32)

(1) ⌜3;0.2⌟ A faint central peak.

(2) ⌜3;U. L.⌟ A faint elevation.

(3) ⌜4/3;0.3w⌟ A low ridge.

Zagut B

Zagut C

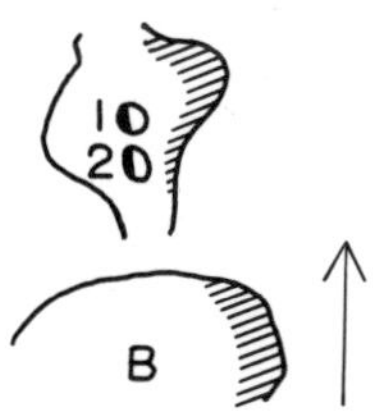

73. Zagut C (+272,-512;24)

(1) ⌜3;U. L.⌟ A very faint central peak.

(2) ⌜4/3;U. L.⌟ An oblong hill on the south wall.

74. Zagut P (+253,-536;15)

(1) ⌜5̇/4;0.2⌟ Very low central mound, somewhat oblong, on the easterly eccentric position.

75. Zagut P, U. F. Containing (+256,-530;41)

(1) ⌜4;1.2w/0.1e⌟ A peak on the easterly eccentric position. It stands on the slope of the wall of Zagut B.

X

1. Behaim (+943,-285;56)

(1) [10;0.5] A large central peak.

2. Behaim B (+935,-278;24)

(1) [5;0.5] A central peak contacting with the north wall of Behaim BA.

3. Bohnenberger (+619,-279;33)

(1) [3;0.7e] (2) [2;0.4e] Twin peaks. The former is higher.

(3) [3.5;0.7w] East rampart of a small crater, whose diameter is about 4 km.

(4) [3.5;0.2e] A low hill connecting with the (1).

(5) [4;U. L.] (6) [3;U. L.] Faint elevations.

Bohnenberger

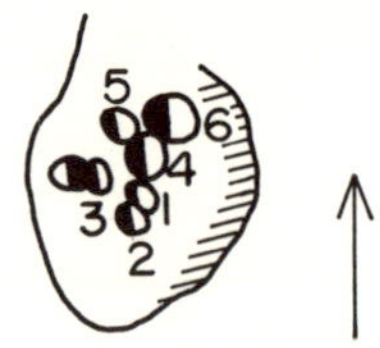

4. Bohnenberger C (+624,-318;16)

(1) [1;U. L.] A very faint mound.

5. Borda (+658,-425;44)

(1) [12;2.0] A large central mountain.

(2) [8;1.5] A hill, somewhat oblong in north-south direction, on the south part of the floor.

Borda

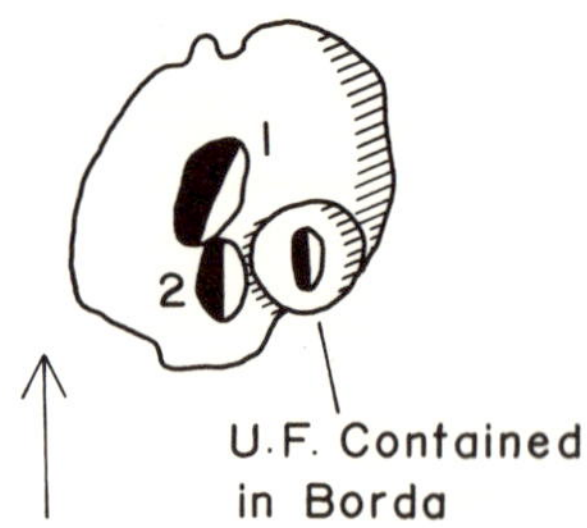

6. Borda, U. F. Contained in (+662,-428;23)

(1) [5/4;<0.5] A very low hill, somewhat oblong in north-south direction.

7. Colombo (+696,-261;78)

(1) [5.5;0.8] The highest peak, having conical form.

(2) [10/5;0.7] An oblong hill.

(3) [1.6;0.2] A very minute peak on whitish region.

(4) [3;0.3] A tiny peak at the bottom of the west wall.

(5) [2;U. L.] (6) [2;U. L.] White patches visible under special illumination.

8. Colombo E, U. F. on the W. of (+643,-273;23)

(1) [4;0.2] A low and broad mound.

Colombo

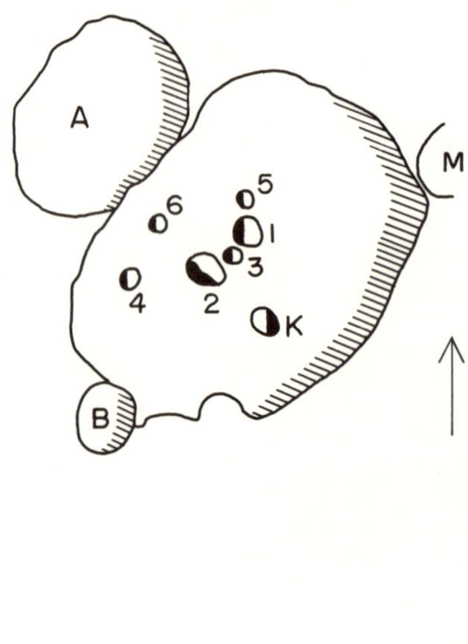

9. Furnerius, U. F. on the W. of (+681,-584;26)

(1) [7;U. L.] Very broad and low elevation.

10. Gaudibert B, U. F. to the W. of (+619,-218;20)

(1) [1;U. L.] On centrally elevated floor, there is a very minute mound.

11. Humboldt, W. (+878,-457;201)

(1) [12;0.3] Two-peaked hill.

(2) [14;0.3] A long ridge.

(3) [10;0.2] Two-peaked white hill.

(4) [5;0.1] A minute hill.

(5) [12;U. L.] A white patch.

12. Humboldt, W., U. F. on the S.-W. of (+845,-506;47)

(1) [15;U. L.] Centrally elevated floor.

13. Lamé (+873,-254;84)

(X)

(1) ⌞9;0.8⌟ A central mountain on rugged floor.

14. Lamé G (+877,-266;26)
    (1) ⌞7/4;U. L.⌟ Very low and broad hill, connected with the central peak of Lamé by a faint dyke.

15. Lohse (+843,-237;42)
    (1) ⌞8;1.3⌟ A distinct central peak.

16. Magelhaens A (+690,-220;30)
    (1) ⌞7;U. L.⌟ Central white patch, probably a faint elevation.

17. McClure (+742,-264;24)
    (1) ⌞3;0.7⌟ A small mound on the northerly eccentric position of the floor.

18. McClure C, U. F. to the N.-W. of (+727,-249;17)
    (1) ⌞13/2;U. L.⌟ A faint and meandering cliff runs on the floor from south to north.

19. Monge (+697,-329;37)
    (1) ⌞5̇;1.0̇⌟ (2) ⌞7;0.7⌟ Two hills stand in north-south line.

20. Palitzsch B (+834,-445;39)
    (1) ⌞5.5;1.5⌟ A distinct central peak.

W. Humboldt

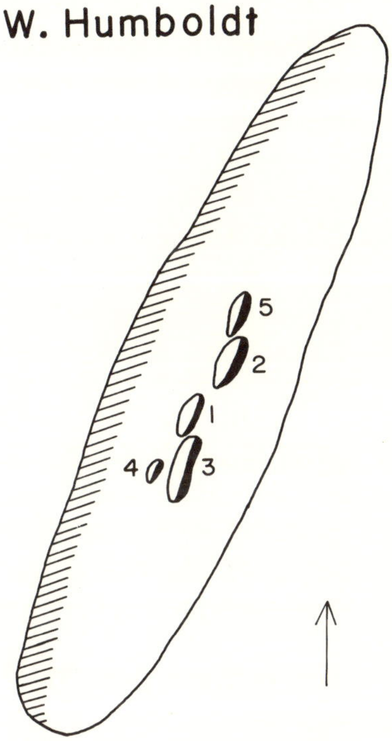

Petavius, Wrottesley

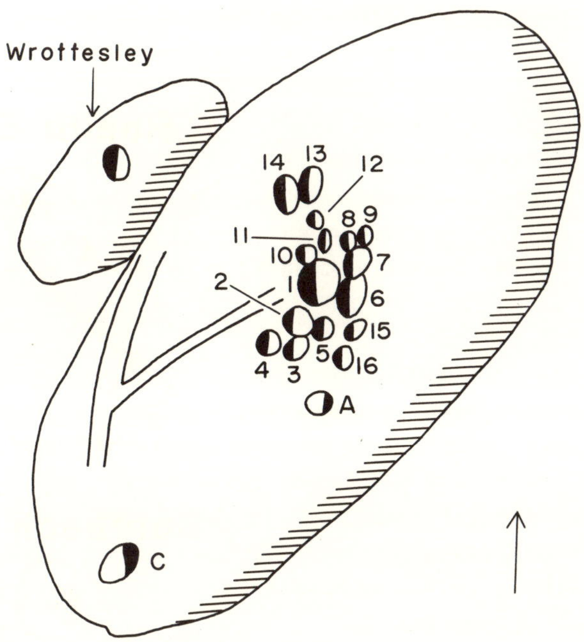

21. Petavius (+786, -427;177)
    (1):α ⌞13;5̇w⌟ The principal peak. A lofty mountain of irregular form, from which a distinct crack runs to the west wall.
    (2) ⌞10̇;5̇w⌟ The second peak.
    (3) ⌞6;1⌟ (4) ⌞4;1⌟ (5) ⌞5̇;0.8e⌟ Mountains on the south of the (2).
    (6) ⌞9;1.2e⌟ A large hill on the east side of the (1).
    (7) ⌞8;1.0e⌟ Lofty mountain of irregular form.
    (8) ⌞5;0.6⌟ (9) ⌞3;0.5̇⌟ (10) ⌞5̇;0.5̇⌟ .(11) ⌞2.5;0.4⌟ (12) ⌞2;0.3⌟ These are forming a group of small hills.
    (13) ⌞8;U. L.⌟ (14) ⌞8;U. L.⌟ Swelled parts of the floor.

Monge

(X)

(15) ⌞4;0.5⌟ (16) ⌞5;0.7⌟ Isolated hills on the east part of the floor.

22. Petavius B (+790,-340;33)
    (1) ⌞7;U. L.⌟ Centrally elevated floor.

Reichenbach

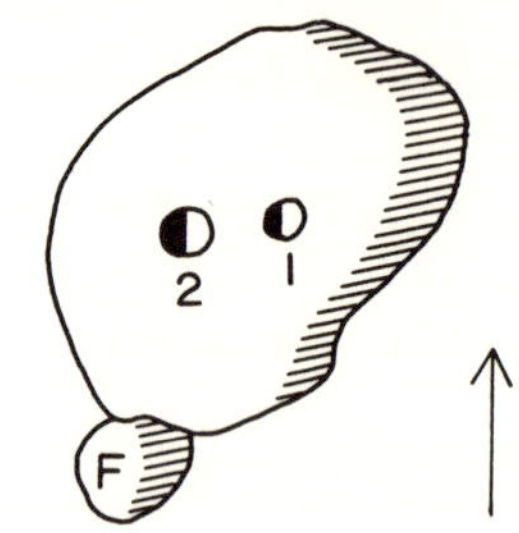

23. Reichenbach (+642,-504;71)
    (1) ⌞4;0.4⌟ A mound on the north-easterly eccentric position of the floor.
    (2) ⌞5̇;U. L.⌟ A very low peak.

24. Reichenbach A (+664,-474;34)
    (1) ⌞10̇/6;0.9⌟ A slightly oblong hill, north part of which is highest.

25. Reichenbach B (+654,-476;44)
    (1) ⌞10̇/4;0.7⌟ A central hill oblong in direction from south-east to north-west.

26. Reichenbach C (+606,-490;28)
    (1) ⌞5̇;0.3⌟ A faint mound at the center of the floor.

Reichenbach D

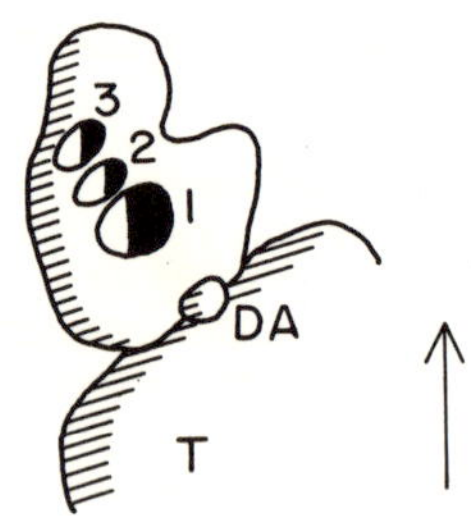

27. Reichenbach D (+621,-471;31)
    (1) ⌞11;1.6⌟ Very broad elevation.
    (2) ⌞10/5;U. L.⌟ (3) ⌞10/5;U. L.⌟ Low parallel hills.

Reichenbach T

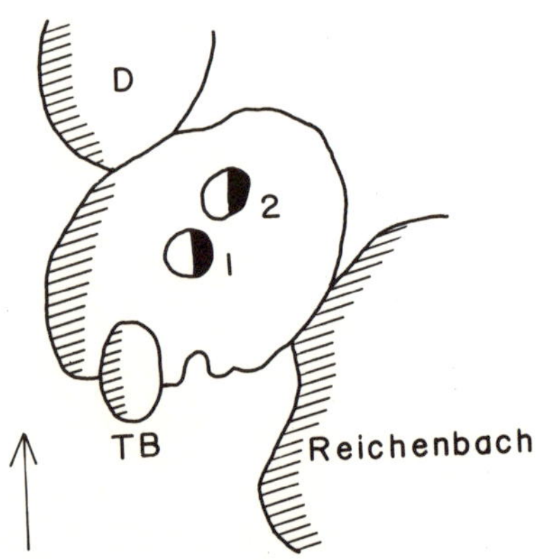

28. Reichenbach T (+624,-490;64)
    (1) ⌞10;1.5⌟ A distinct peak.
    (2) ⌞7;1.0⌟ A lower peak on the north part of the floor.

29. Reichenbach Z (+610,-529;15)
    (1) ⌞2.5;0.8⌟ A small central mound.

Rheita E

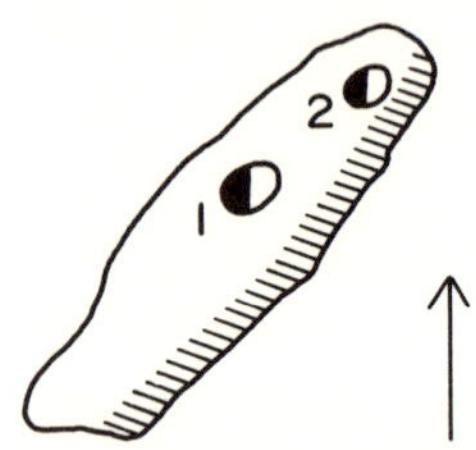

30. Rheita E (+625,-562;66/32)
    (1) ⌞8;1.0⌟ A round mound, situated on northerly eccentric position.
    (2) ⌞5̇;1.0⌟ A round mound, situated on the bottom of the north wall.

Santbech

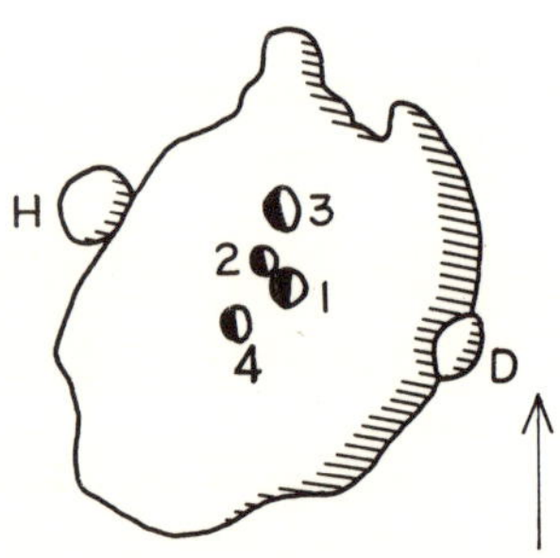

31. Santbech (+650,-356;64)
    (1):β ⌞5̇;0.8⌟ (2) ⌞3;0.6⌟ The (1) is a cone-formed peak, accompanied by the lower hill (2) on the north-west of the (1).
    (3) ⌞7;U. L.⌟ Very low and broad elevation.
    (4) ⌞3;U. L.⌟ A faint mound.

Santbech A

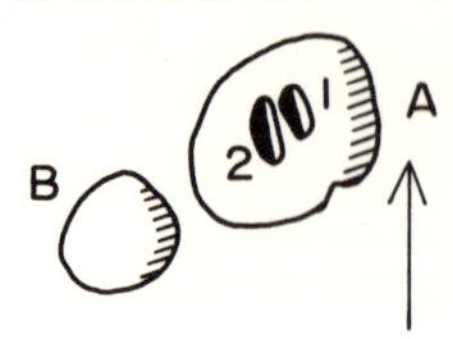

32. Santbech A (+615,-411;25)
    (1) ⌞5.0/2.5;0.5⌟ (2) ⌞5.0/2.5;0.4⌟ Both are oblong mounds, parallel to each other.

33. Snellius (+720,-490;83)

(1) ⌞8;1.1e/1.7w⌟ A cone-like, but gentle central hill. The south part of the floor is very rugged, and contains a crack.

34. Snellius A (+716,-460;37)

(1) ⌞7;<0.5⌟ Very faintly elevated floor.

35. Snellius A, U. F. on the N. of (+718,-445;21)

(1) ⌞4;0.8⌟ (2) ⌞3;0.7⌟ (3) ⌞4;0.8⌟ Small mounds, among which the (2) is most faint.

(4) ⌞3;0.6⌟ A faint mound.

(5) ⌞4;<0.5w⌟ A whitish mound, situated on the south wall.

U. F. on the North of Snellius A

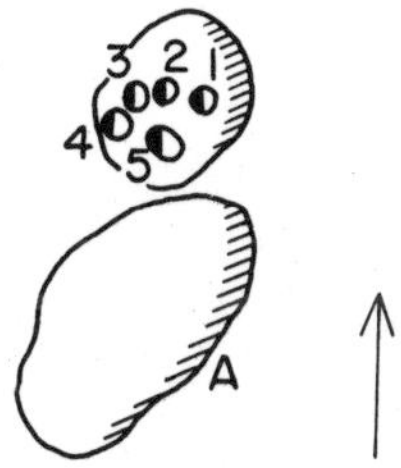

36. Snellius D, U. F. on the E. of (+692,-479;25)

(1) ⌞5;0.8⌟ An oblong central mound.

(2) ⌞5;0.6⌟ A gentle hill on the south of the (1).

U. F. on the East of Snellius D

Stevinus

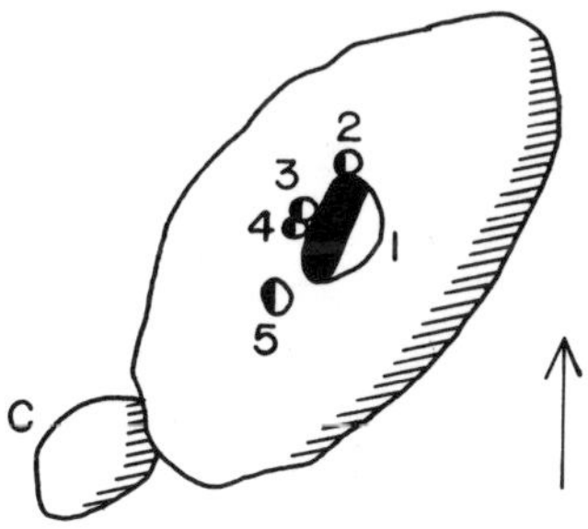

37. Stevinus (+684,-537;74)

(1) ⌞15;1.6e/2.4w⌟ A large, ridge-like mountain.

(2) ⌞2;0.3⌟ (3) ⌞2;0.3⌟ (4) ⌞2;0.3⌟ Tiny mounds around the (1).

(5) ⌞4;1.2⌟ A distinct mound.

38. Stevinus B (+680,-517;20)

(1) ⌞5;U. L.⌟ A faintly elevated floor.

39. Stevinus D (+637,-571;22)

(1) ⌞3;<0.6⌟ A faint central mound.

(2) ⌞3;<0.6⌟ A faint mound situated on the south of the (1).

Stevinus D

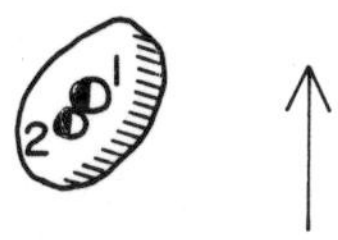

40. Wrottesley (+765,-405;57)

(1) ⌞5.5;1.4⌟ A distinct central mountain, having a rather irregular form.

XI

1. Apollonius F (+862,+098;16)
   (1) ⌞2.5;U. L.⌟ On the west wall, there is a gentle mound.
2. Auzout (+885,+178;33)
   (1) ⌞5̇;<1.2⌟ A faint elevation, probably the true central peak.
   (2) ⌞9;<1.2⌟ A broad hill, having a little irregular form.
3. da Vinci (+698,+160;31)
   (1) ⌞7;2⌟ A round central mound.
   (2) ⌞6;0.8⌟ (3) ⌞5̇;0.8⌟ (4) ⌞5̇;1.0̇⌟ Elevated parts of the rugged floor.
   (5) ⌞3;0.8⌟ (6) ⌞10/4;1.5̇⌟ Collapsed parts of the east wall.
4. Gaudibert (+602,-190;29)
   (1) ⌞11/9;2.5⌟ (2) ⌞11/6;2.0⌟ Two ridge-like mountain blocks lie on the floor, among which the (1) is larger. The hill (2) lies on the east of the (1).
5. Gilbert, U. F. of the S. of (+967,-086;57)
   (1) ⌞4;U. L.⌟ A low central peak.
6. Gilbert J (+952,-075;38)
   (1) ⌞4;U. L.⌟ A low central peak.
7. Goclenius (+696,-174;66/54)
   A distinct crack, named Rimae Goclenius II, passes through this crater.
   (1) ⌞3;0.4⌟ A small and cone-formed central peak, standing on the crack. The east half part of the floor is relatively smooth.
   (2) ⌞2;0.3⌟ An isolated peak.
   (3) ⌞5;0.5⌟ (4) ⌞3;0.2⌟ Two dykes which are connected with each other, and form the east bank of the crack.
   (5) ⌞5;0.3⌟ Opposite bank of the (3) and (4).
   (6) ⌞3.5;U. L.⌟ Broad and faint elevation.
   (7) ⌞3;U. L.⌟ (8) ⌞3;U. L.⌟ White patches.
   (9) ⌞8;U. L.⌟ Broad elevation.
   (10) ⌞5;0.5⌟ (11) ⌞3;0.3⌟ (12) ⌞3;0.3⌟ Collapsed parts of the north wall.
   (13) ⌞5;0.5⌟ A very faint peak.
8. Gutenberg (+652,-151;71) A crack, named Rimae Gutenberg I, runs on the floor.
   (1) ⌞10/5̇;1.0̇⌟ A ridge-formed central peak.
   (2) ⌞6;0.3⌟ A small mountain on the north side of the (1).
   (3) ⌞3.5;0.2⌟ A faint dark hill on the south side of the (1).
   (4) ⌞8/5;0.2⌟ (5) ⌞11/5̇;1.0̇⌟ Oblong hills, which are parallel to each other.
   (6) ⌞2;0.2⌟ (7) ⌞2;0.2⌟ (8) ⌞2;0.2⌟ (9) ⌞2;0.2⌟ A group of small mounds between the crack and the mountain (2), among which the (9) is specially white.
   (10) ⌞4;0.3w⌟ (11) ⌞3;0.3w⌟ (12) ⌞4;0.3w⌟ Three mounds which form east bank of the crack.

da Vinci

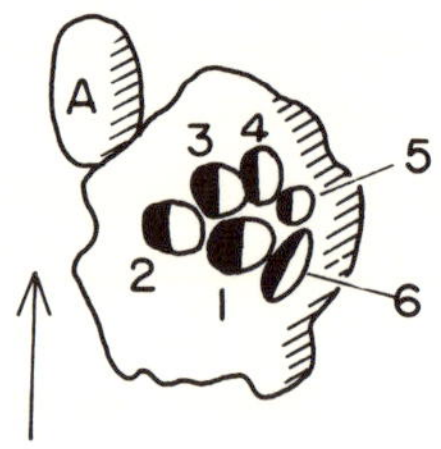

Auzout

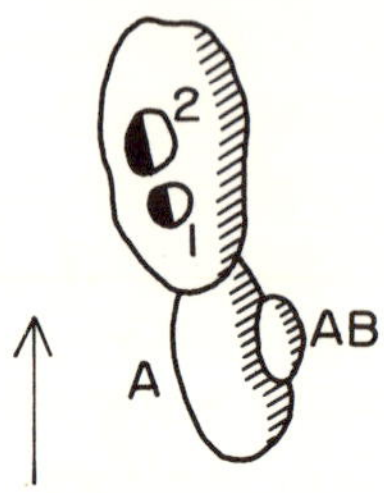

Gaudibert

Goclenius

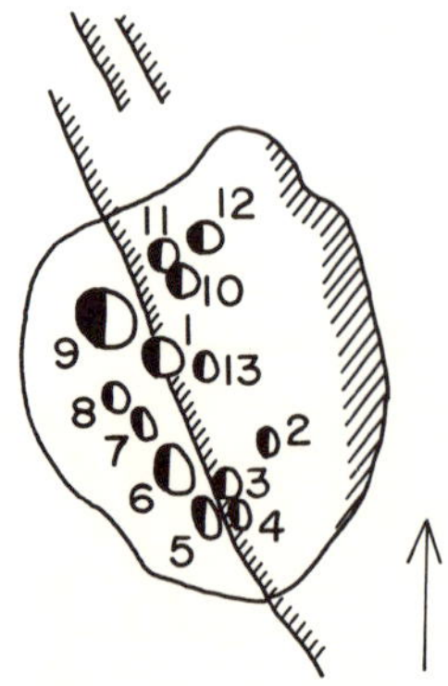

(XI)

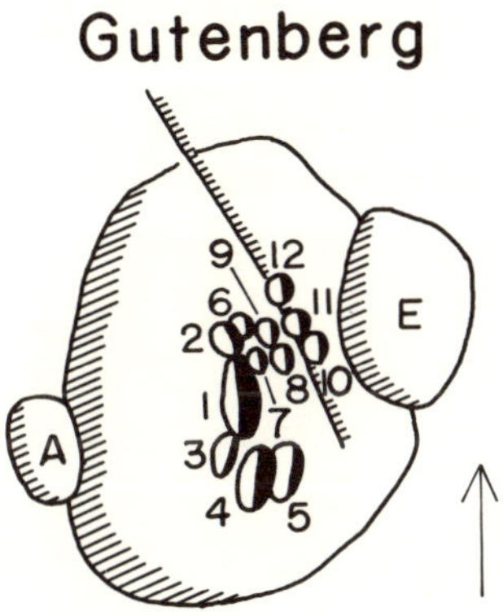

9. Gutenberg C (+649,-175;46)

(1) ⌞15/8;1.2⌟ The most distinct ridge.

(2) ⌞5.0;0.5⌟ A faint peak, probably a part of the south wall of Gutenberg.

(3) ⌞4;U. L.⌟ A broad swelling part of the floor.

(4) ⌞10/6;U. L.⌟ (5) ⌞8.5;U. L.⌟ The (4) is a faint mound, and connected with the faint ridge (5).

(6) ⌞7;U. L.⌟ A swelling part of the floor.

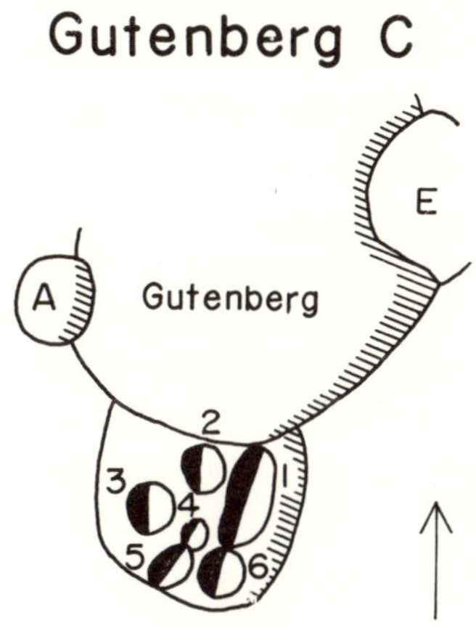

10. Gutenberg G (+639,-106;31)

(1) ⌞4;1.1⌟ A typical central peak.

11. Kapteyn (+927,-162;49)

(1) ⌞4;<1⌟ (2) ⌞4;<1⌟ Twin central peaks. The northern component, (1), is higher than the southern one, (2).

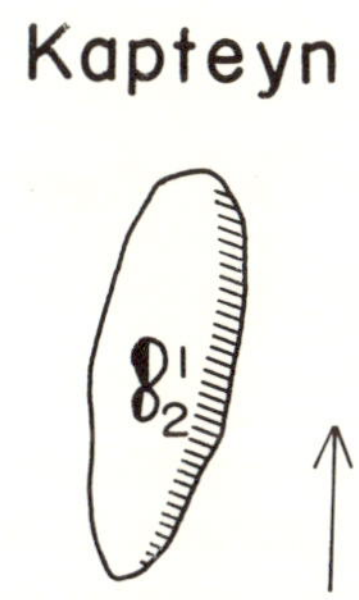

12. Kapteyn E (+924,-153;31)

(1) ⌞5;1.5⌟ Rather high central peak.

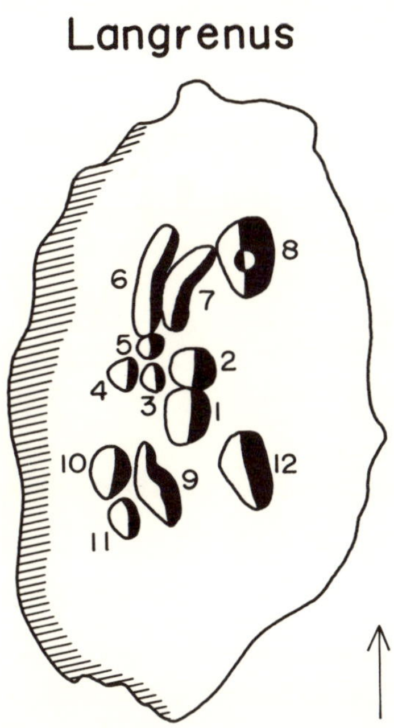

13. Langrenus (+863,-155;132)

(1):α ⌞11;3.5⌟ (2):β ⌞11;3.0⌟ The main peak comprises two bodies, (1) and (2). The southern component (1) is slightly higher than the northern one, (2). Both have somewhat irregular forms.

(3) ⌞5.5;U. L.⌟ (4) ⌞7;U. L.⌟ (5) ⌞6;U. L.⌟ Round faint mounds on the west bottom of the (2).

(6) ⌞28/8;U. L.⌟ (7) ⌞20/10;U. L.⌟ Long faint dykes, which seem as traces of terrace structure.

(8) ⌞15;U. L.⌟ A broad elevation having a depression on its center.

(9) ⌞20/8;U. L.⌟ (10) ⌞7;U. L.⌟ (11) ⌞10;U. L.⌟ White patches.

(12) ⌞10;U. L.⌟ A faint elevation.

14. Langrenus A (+906,-184;43)

(1) ⌞5;<0.8⌟ Low central peak.

(2) ⌞4;U. L.⌟ A peak standing on the northerly eccentric position of the floor, and lower than the (1).

15. Langrenus K (+841,-101;29)

(XI)

(1) ⌊7;0.9⌋ A low mound.

(2) ⌊4;0.6⌋ From north-west wall, a dyke, (2), runs towards the (1).

Langrenus A

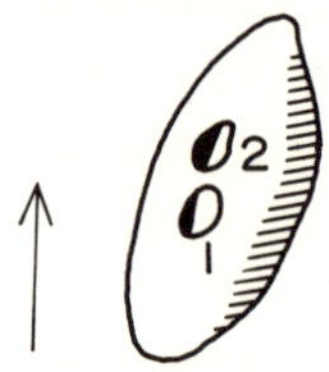

Langrenus K

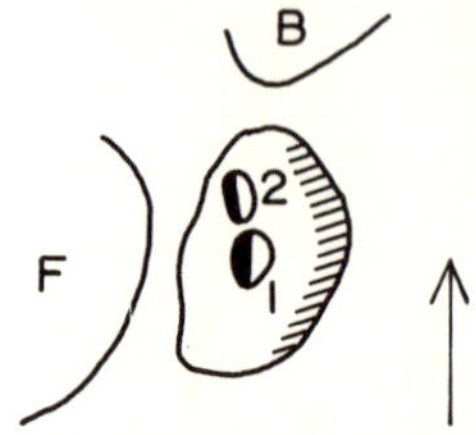

16. La Pérouse (+956,-185;78)

(1) ⌊8;2⌋ Main peak having a conical form.

(2) ⌊4;1⌋ A faint hill.

(3) ⌊5;2⌋ A peak standing on the south-easterly eccentric position of the floor.

U.F. to the South of Lubbock J

17. Lubbock J, U. F. to the S. of (+649,-046;28/21)

(1) ⌊3;0.5⌋ A round mound.

(2) ⌊1.5;U. L.⌋ A very tiny mound.

(3) ⌊6/2.5;0.3⌋ An oblong dyke, of which northern part is higher.

La Pérouse

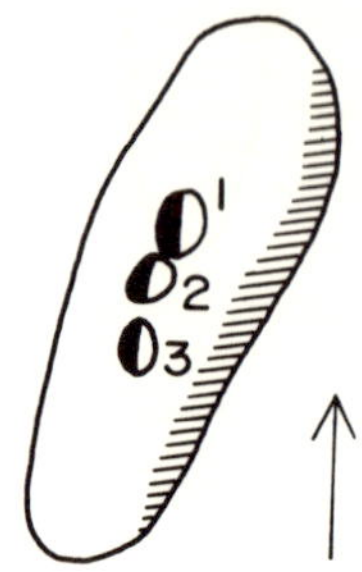

18. Lubbock M (+625,-006;19)

(1) ⌊5;U. L.⌋ Very faintly elevated floor.

19. Lubbock P, U. F. to the W. of (+622,-046;28)

(1) ⌊6;0.7⌋ On the south-westerly eccentric position of the floor, there is a mountain block, of which south-east part is depressed.

20. Maclaurin (+927,-033;50)

(1) ⌊4;<0.6⌋ A tiny central hill.

21. Maclaurin A (+923,-053;29)

(1) ⌊3;U. L.⌋ The central peak is very indistinct under ordinary illumination. Only at the full moon, it can be recognized as a white spot.

22. Maclaurin L (+949,-024;30)

(1) ⌊6;U. L.⌋ On the south-west part of the floor, there is a faint elevation.

23. Maclaurin L, U. F. to the N. of (+950,-001;22)

(1) ⌊4;U. L.⌋ Very faintly centrally elevated floor.

24. Maclaurin M (+930,-084;42)

(1) ⌊5;<0.5⌋ Very faint central peak, which is also a part of another unnamed crater.

25. Maclaurin O (+925,-005;53)

(1) ⌊5.5;U. L.⌋ Under the ordinary illumination, a black patch is seen at the center of the floor. It looks like a white patch near the full moon.

26. Maclaurin P, U. F. on the E. of (+927,-100;45)

(1) ⌊45/10;U. L.⌋ This formation is not a crater, but one of parallel structures from Maclaurin B to Maclaurin F. A faint, long ridge runs through this formation.

27. Neper (+983,+155;142)

(1) ⌊4;2⌋ (2) ⌊5;1⌋ Twin mountain block stands on the center of the floor.

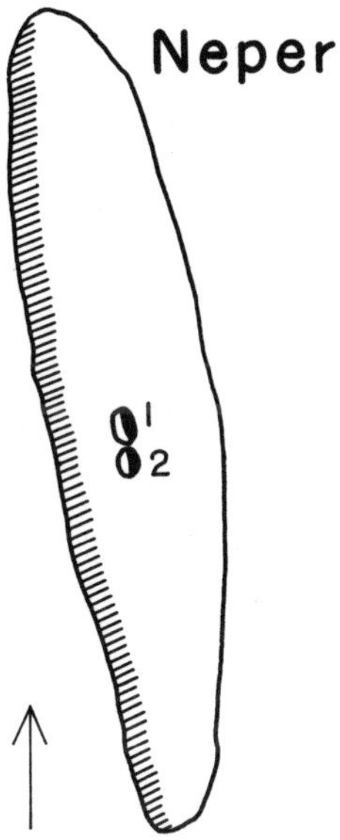

The northern component, (1), is higher than the southern one, (2).

28. Schubert (+987,+047;54)
    (1) ⌞5;U. L.⌟ (2) ⌞5;U. L.⌟ Both objects are very faint central elevations.

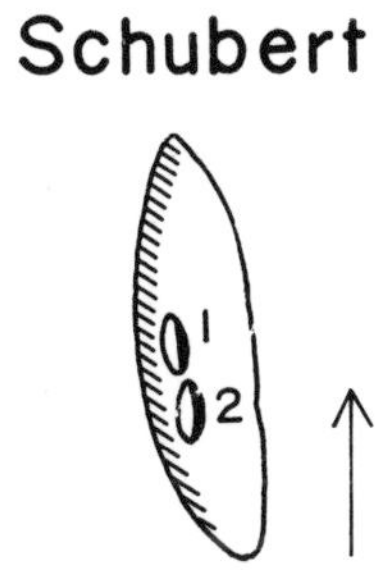

29. Secchi (+688,+042;24)
    (1) ⌞5̇;<0.5̇⌟ A broad elevated part of the floor.
    (2) ⌞1.5;U. L.⌟ (3) ⌞2;U. L.⌟ (4) ⌞1.5;U. L.⌟ Faint mounds.
30. Secchi, U. F. on the S.-W. of (+684,+033;16)
    (1) ⌞4;<0.4⌟ Centrally elevated floor.

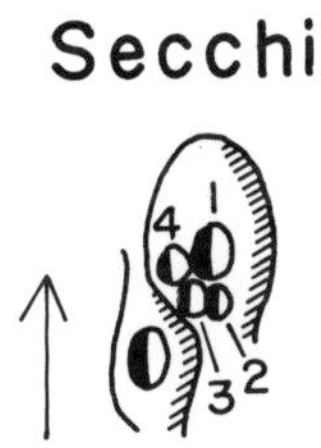

31. Taruntius (+722,+098;56)
    (1) ⌞6;1.9w/0.7e⌟ A cone-formed central crater. There is a concentric pattern on the floor.
32. Webb (+865,-016;23)
    (1) ⌞5;U. L.⌟ Very faintly centrally elevated floor.

XII

1. Bernouilli (+714,+574;47)
   (1) ⌞8;1.4⌟ A peak which contains two mountain blocks.
2. Burckhardt (+714,+517;57)
   (1) ⌞10/4;0.8⌟ An oblong hill on the west part of the floor.
   (2) ⌞12;0.6⌟ A low mountain block on the south part of the floor.

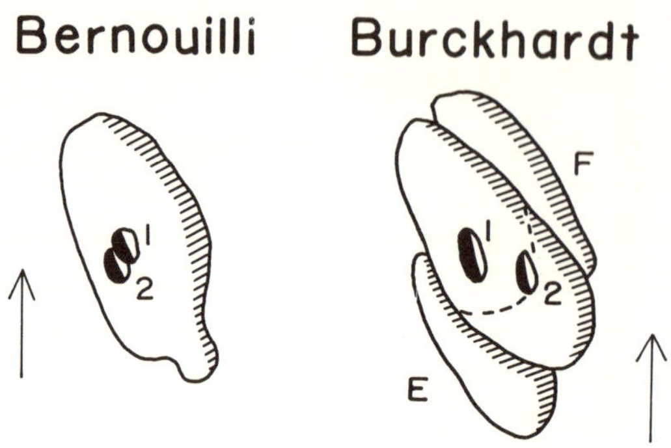

3. Cleomedes (+730,+465;126)
   (1):α ⌞10;1.3⌟ A ridge-like mountain.
   (2) ⌞6;0.5⌟ (3) ⌞5;0.4⌟ Attached hills of the (1).
4. Cleomedes D, U. F. Containing (+771,+485;40)
   (1) ⌞5̇;1.0⌟ A round mound, which is usually regarded as a part of south wall of Cleomedes D.
5. Cleomedes E (+714,+479;21)
   (1) ⌞8;U. L.⌟ A centrally elevated floor.

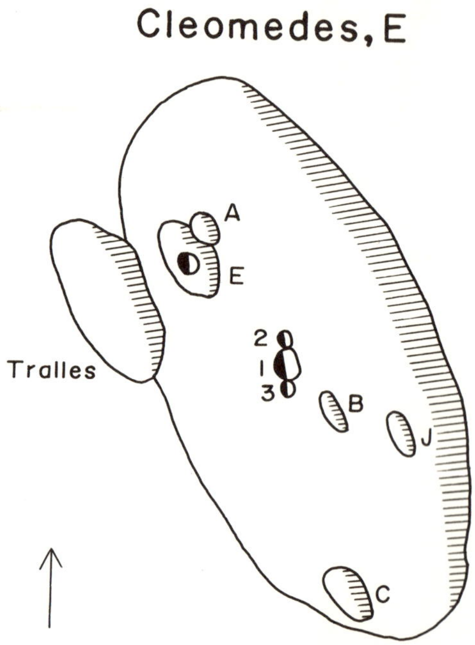

6. Debes (+687,+493;31)
   (1) ⌞4;U. L.⌟ A very faint mound at the bottom of the west wall.
7. Delmotte (+772,+456;33)
   (1) ⌞3;<0.8⌟ A very faint oblong mound, situated on the north-easterly eccentric position.
   (2) ⌞5̇;U. L.⌟ An elevated part of the floor.
8. Gauss (+795,+587;177)
   (1) ⌞13;0.5̇⌟ The most distinct peak, somewhat whitish, of which north part is highest.
   (2) ⌞4;0.2⌟ (3) ⌞6;0.3⌟ (4) ⌞2;0.1⌟ (5) ⌞8;0.4⌟
   (6) ⌞5;0.4⌟ The peaks from (1) to (6) form a mountain ring, concentric with the crater wall.
   (7) ⌞7;0.5̇e⌟ (8) ⌞18;0.7e⌟ Mountain blocks on the north-east part of the floor.
   (9) ⌞13;U. L.⌟ (10) ⌞7;U. L.⌟ (11) ⌞10;U. L.⌟ Hills on the rugged part of the floor.

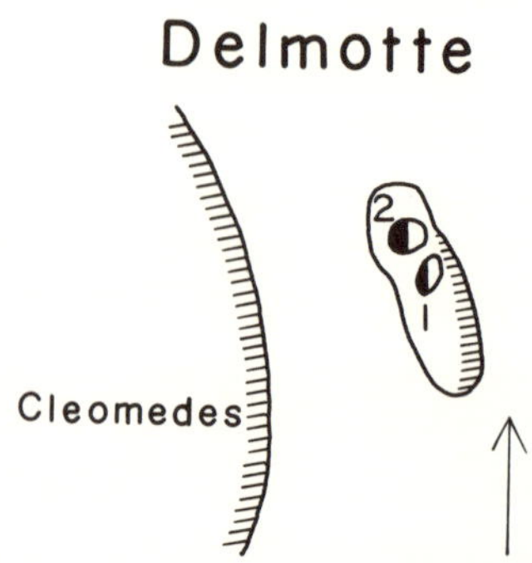

9. Gauss, U. F. to the S.-E. of (+872,+489;102)
   This formation was formerly called Rayleigh.
   (1) ⌞14;3⌟ A large mountain on the western part of the floor.
   (2) ⌞30;2⌟ A concentric crater with the present formation, whose diameter is

Gauss

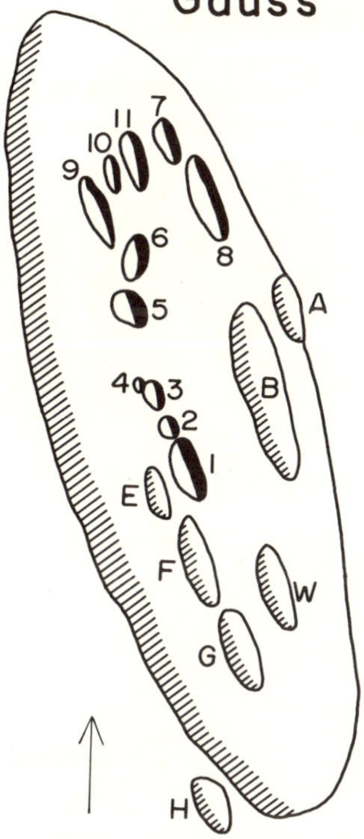

U.F. to the S.E. of Gauss

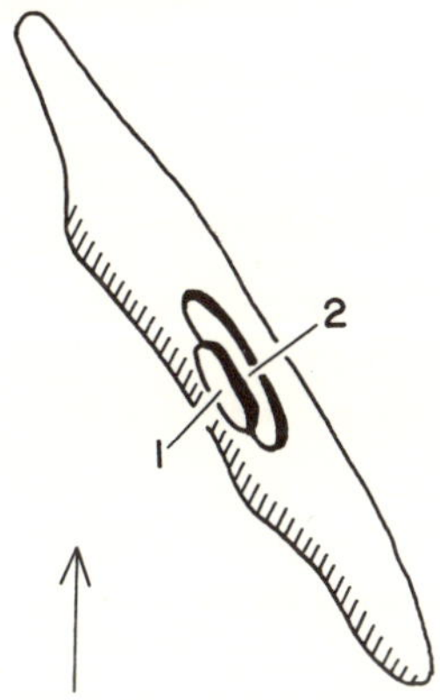

is about 15 km.

10. Geminus (+689,+566;85)
   (1) ⌞8;1.0⌟
      The central peak is twin-peaked, of which southern one is higher.

11. Geminus, U. F. on the S. of (+711,+543;20)
   (1) ⌞6.5;0.6⌟ A round mound.

12. Hahn (+820,+519;84)
   (1) ⌞20;3.8⌟ A large central peak, having a conical form. (Confirmed by Lunar orbiter IV Photographs.)

13. Hansen (+926,+242;40)
   (1) ⌞8; U. L.⌟ A round low mound at the center of bowl-formed floor. (According to the Lunar Orbiter IV Photographs, it is collapsed form of a mountain.)

14. Macrobius (+671,+363;64)
   (1) ⌞14;1.6⌟ The central peak, having a conical form.
   (2) ⌞10;0.8⌟ A low hill on the north-west of the (1).
   (3) ⌞4;U. L.⌟ (4) ⌞4;U. L.⌟ (5) ⌞4;U. L.⌟ Faint elevations on the north part of the floor.

Macrobius

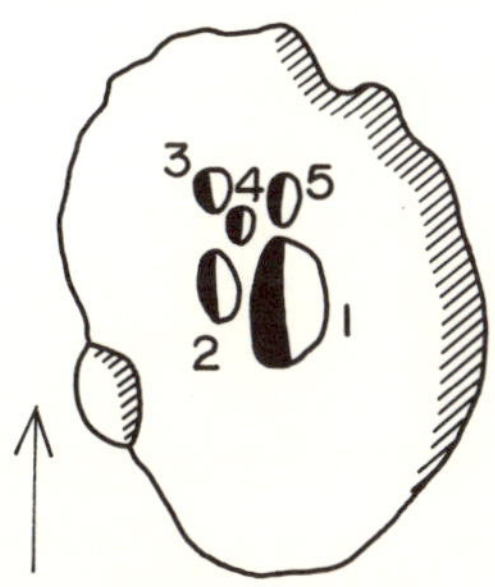

15. Messala A (+648,+596;26)
   (1) ⌞3;<0.5⌟ A low hill at the north-west border of the floor.

16. Newcomb J (+612,+481;23)
   (1) ⌞6;<1.0⌟ A broad hill, having somewhat oblong form.

17. Newcomb P (+643,+491;87/65)
   (1) ⌞6;0.4⌟ The most distinct hill on the north-westerly eccentric position of the floor.
   (2) ⌞4;0.1⌟ Small attached hill.
   (3) ⌞4;0.2⌟ A mound at the bottom of the west wall.
   (4) ⌞4;0.3⌟ A mound at the bottom of the east wall.
   (5) ⌞2.5;U. L.⌟ (6) ⌞2.5;U. L.⌟ (7) ⌞2.5;U. L.⌟ Low peaks standing on a faint scarp.
   (8) ⌞2.5;U. L.⌟ (9) ⌞2;U. L.⌟ Small peaks at the north-west part of the floor.
   (10) ⌞2.5;U. L.⌟ A mound on the south part of the floor.

18. Peirce (+761,+313;18)

(XII)

(1) ⌜4/3;<0.9⌟ A somewhat oblong hill on the south-westerly eccentric position of the floor.

Newcomb P

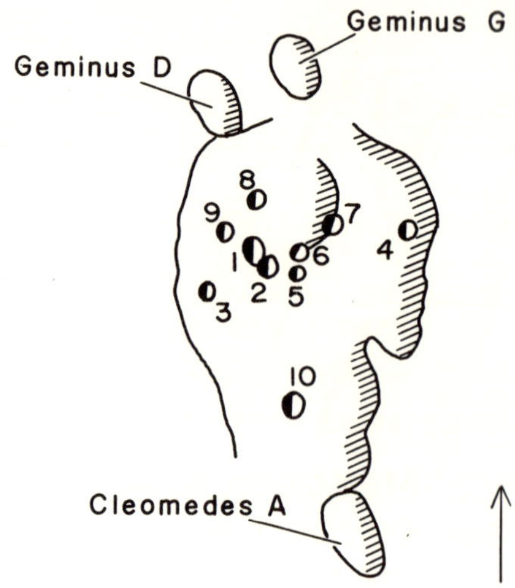

19. Peirce C,U. F. on the N.-W. of
(+719,+329;15)
(1) ⌜3.5;0.7⌟ A round central mound.
20. Plutarch (+897,+408;68)
(1) ⌜10;2⌟ A large central peak. (Confirmed by the Lunar Orbiter IV Photograph.)

Proclus

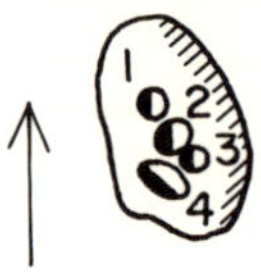

21. Proclus (+702,+278;28)
(1) ⌜3;<0.6⌟ A round mound.
(2) ⌜3;U. L.⌟ (3) ⌜3;1w⌟ Faint mound.
(4) ⌜7/3;1w⌟ Collapsed part of the south-west wall.
22. Seneca (+878,+454;63)
(1) ⌜5;U. L.⌟ A faint hill on the southerly eccentric position of the floor.

Tralles

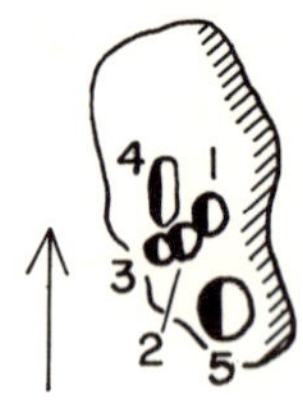

23. Tralles (+700,+476;43)
(1) ⌜6;1.5⌟ (2) ⌜4;1.0⌟ (3) ⌜4;1.0⌟ These three mounds stand in a line. The (1) is the highest.
(4) ⌜5;1.5⌟ Another mound at the bottom of the west wall.
(5) ⌜7;U. L.⌟ Elevated part of the floor.
24. Yerkes (+759,+252;36)
(1) ⌜4;<0.3⌟ Very faint mound at the center of the floor.

XIII

1. Gauss, U. F. to the N.-E. of (+770,+635;133)
This formation was formerly called Riemann.
(1) ⌞12;1⌝ The most distinct peak on the north-easterly eccentric position of the floor.
(2) ⌞10;U. L.⌝ (3) ⌞12;U. L.⌝ (4) ⌞13;U. L.⌝ Very faint elevations.
(5) ⌞15;U. L.⌝ A faint ridge at the bottom of the wall.

U.F. to the N.E. of Gauss

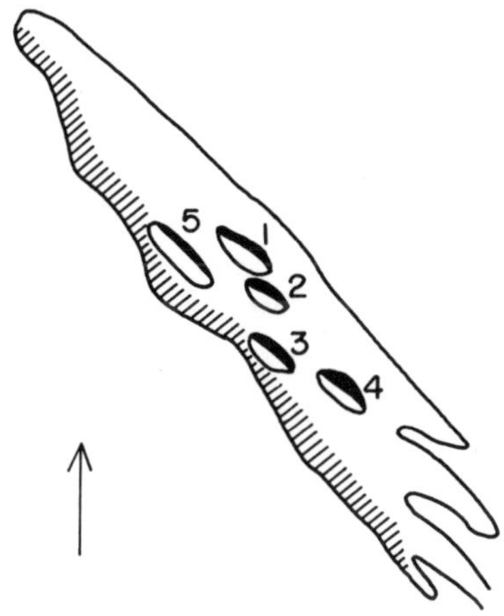

2. Mercurius (+629,+726;68)
(1) ⌞3;U. L.⌝ A white patch on somewhat southerly eccentric position of the floor.

3. Mercurius A (+642,+743;20)

Mercurius A
(Based on the Lunar Orbiter IV Photographs)

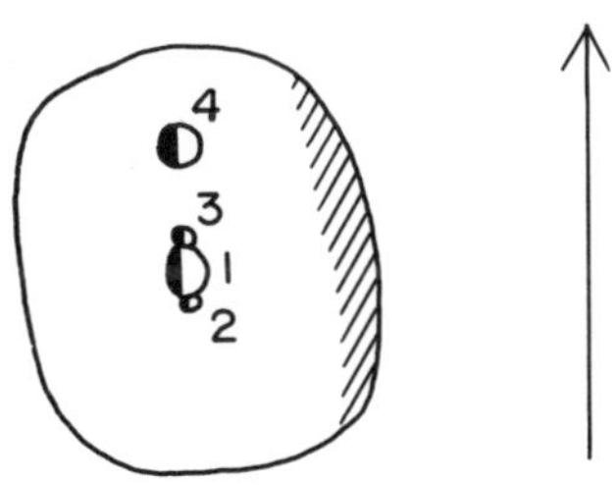

(1) ⌞3;0.1⌝ (2) ⌞2;U. L.⌝ (3) ⌞2;U. L.⌝
The (1) is a central peak accompanied by small mounds (2) and (3) on its south and north sides respectively.
(4) ⌞3;U. L.⌝ A mound with rugged surface, according to the Lunar Orbiter IV Photographs.

Messala

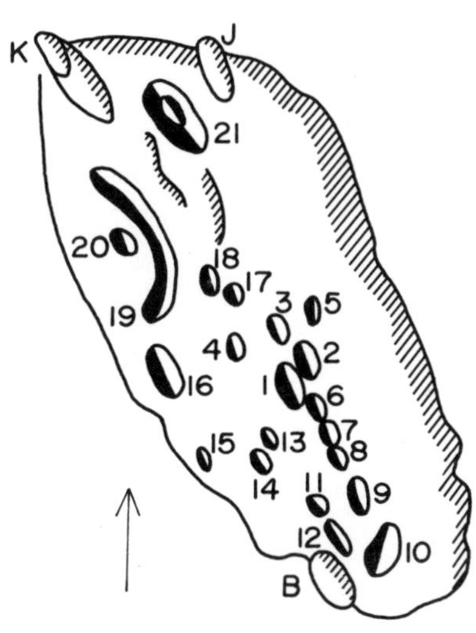

4. Messala (+670,+632;124)
(1) ⌞14;1.0⌝ (2) ⌞8;0.5⌝ (3) ⌞4;0.3⌝ Higher parts of rims of three shallow craters.
(4) ⌞4;0.3⌝ (5) ⌞4;0.3⌝ Faint mounds.
(6) ⌞5;0.4⌝ (7) ⌞4;0.4⌝ (8) ⌞4;0.4⌝ (9) ⌞3;0.3⌝ (11) ⌞3;0.2⌝ (12) ⌞7;0.2⌝ These hills form a faint mountain ring on the southern part of the floor.
(10) ⌞18/6;1⌝ A large hill which is a collapsed part of the south wall.
(13) ⌞3;0.4⌝ (14) ⌞3;0.3⌝ Rather distinct mounds.
(15) ⌞3;0.4⌝ A distinct peak on the bottom of the west wall.
(16) ⌞11;U. L.⌝ A broad faint elevation.
(17) ⌞2;U. L.⌝ (18) ⌞2;U. L.⌝ Faint mounds.
(19) ⌞30/3;0.8(highest part)⌝ (20) ⌞4;0.4⌝ The (19) is rim of a ghost crater, whose central peak is the (20).
(21) ⌞15;0.4⌝ A crater on a broad elevation.

5. Messala D (+704,+650;28)
(1) ⌞4;0.9⌝ Probably a higher part of rim of a small crater on the north-

western part of the floor.

6. Shuckburgh A (+602,+683;19)
   (1) [6;1] A large mountain lies through the north-east wall.
7. Zeno (+674,+709;65)
   (1) [10;0.16] The central peak having a pit, whose diameter is 3 km, on its northern slope.
   (2) [5;0.08] Minute oblong hill, according to the Lunar Orbiter IV Photograph.
8. Zeno A (+670,+700;44)
   (1) [5;0.8] A hill on the northerly eccentric position.
9. Zeno B (+680,+695;37)
   (1) [7;0.3] A central mountain at the center of the crater, according to the Lunar Orbiter IV Photograph.

Zeno, A, B

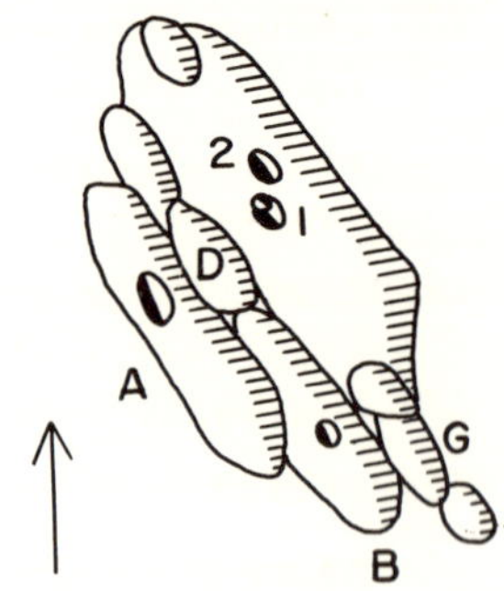

1. Arnold A (+232,+932;57)
   (1) [6;0.5̇] A low peak, which is cone-formed, but somewhat oblong in north-west to south-east directions, stands on the easterly eccentric position of the floor.
2. Arnold K (+222,+944;29)
   (1) [6;0.5̇] A round distinct mound stands on the southerly eccentric position of the floor. In the Arizona Catalogue,the limit of Arnold K is so narrowly estimated that this peak is regarded as a part of west wall.
3. Atlas (+481,+727;87)
   (1) [7;0.9] The most distinct, cone-formed peak, standing on the westerly eccentric position of the floor.
   (2) [4.5;0.6] Low mound, which may be the true central peak.
   (3) [2;0.2] Tiny mouhd on the south of the (1).
   (4) [3;0.2] (5) [5̇;0.5̇] (6) [2;0.2] Three mounds standing in a line, among which the (5) is the largest one.

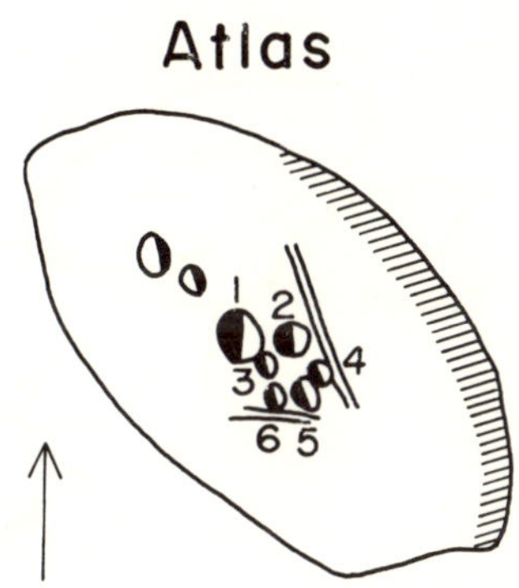

4. Atlas E (+447,+750;58)
   (1) [7;0.7] A cone-formed central peak.

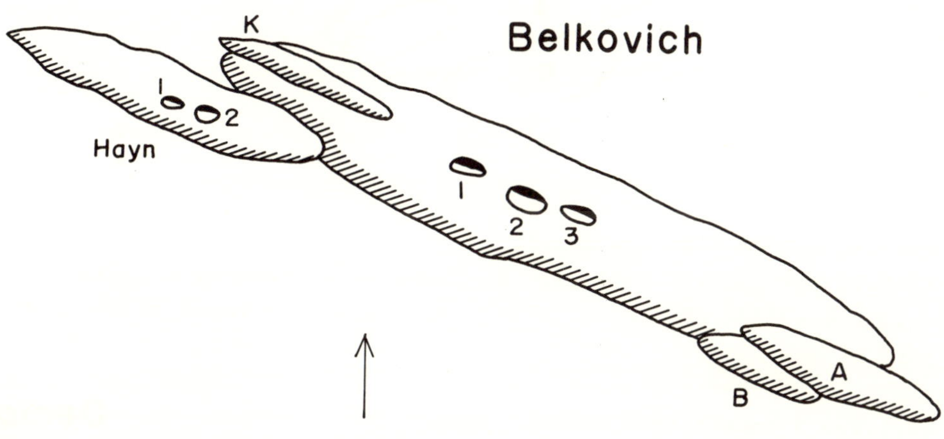

5. Belkovich (+479,+877;198)
   (1) [7;1.5] The most distinguished peak on the westerly eccentric position of the floor.
   (2) [9;1.0] A peak lower than the (1), on the center of the floor. It may be the true central peak.
   (3) [7;1.0] The most faint peak.
6. Bürg (+334,+708;40)
   (1) [10;1.2] A large central mountain.
   (2) [4;0.4] An attached hill on the south bottom of the (1).
   (3) [1.5;0.1] Very tiny mound on the north-west bottom of the (1).
7. Cepheus (+543,+653;40)

Bürg

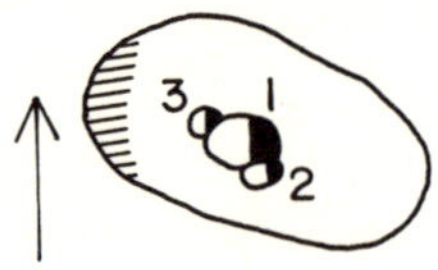

(1) ⌞5;0.4⌟ A mountain having a form of sector.

(2) ⌞3.5;0.3⌟ A round mound to the west of the (1).

8. <u>de la Rue</u> (+410,+858;136)

Most of the objects may be remnants of old craters.

Cepheus

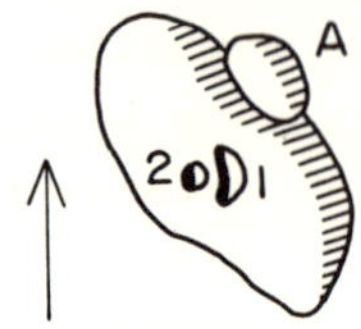

de la Rue

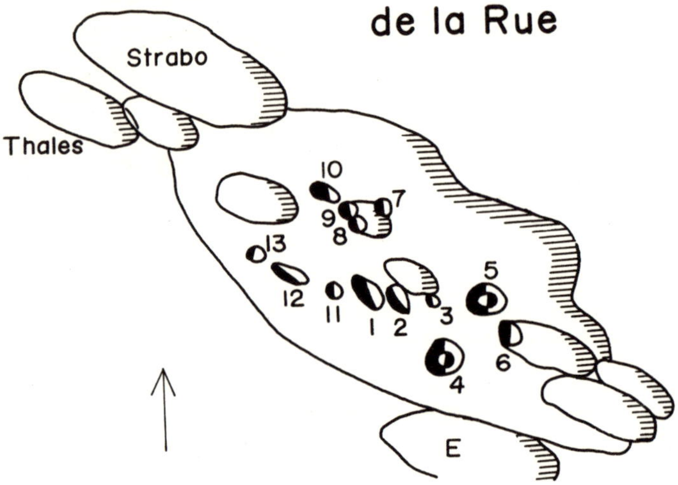

(1) ⌞10;1.0⌟ (2) ⌞7;0.5⌟ Large ridges, parallel to each other, standing on the westerly eccentric position.

(3) ⌞3;U. L.⌟ Very faint mound on the south-east bottom of the crater de la Rue J.

(4) ⌞7;U. L.⌟ (5) ⌞10;U. L.⌟ Small craters with elevated walls.

(6) ⌞8;0.5⌟ A ridge, which may be a part of collapsed crater.

(7) ⌞5;0.4⌟ Rather distinct mound.

(8) ⌞2;0.3⌟ (9) ⌞2;0.3⌟ (10) ⌞4;0.2⌟ Three hills forming a train.

(11) ⌞3;U. L.⌟ (12) ⌞8/3;U. L.⌟ (13) ⌞5;U. L.⌟ Very faint hills on the west part of the floor.

9. <u>Democritus</u> (+267,+886;39)

(1) ⌞7;1.2e⌟ A cone-formed central peak.

(2) ⌞2;U. L.⌟ A small, faint and whitish mound on the west bottom of the (1).

Democritus

10. <u>Endymion C</u> (+457,+852;32)

(1) ⌞5;1⌟ A peak contacting with the south wall.

Endymion J

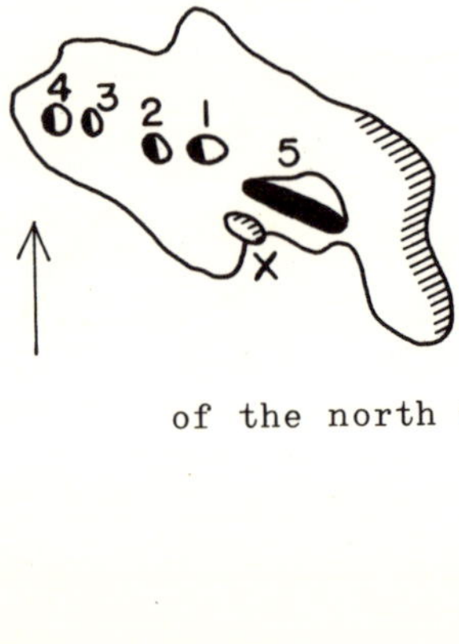

11. <u>Endymion J</u> (+460,+804;67)

(1) ⌞4;0.3⌟ A peak on the northerly eccentric part of the floor.

(2) ⌞2;0.1⌟ (3) ⌞3;0.2⌟ Linearments which are parallel to each other. The (2) is very faint.

(4) ⌞3;0.2⌟ A small, but distinct mound on the bottom of the north wall.

(5) ⌞15/8;0.5w⌟ A broad ridge, which may be remnant of old formation.

Eudoxus

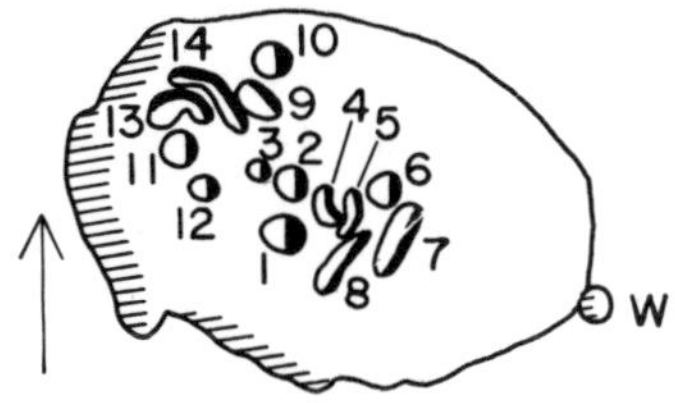

12. Eudoxus (+201,+698;67)

   (1) ⌞5̇;0.6⌟ The most distinct mountain standing on the westerly eccentric position of the floor. This is a twin peak arranged in north-south direction.

   (2) ⌞3;0.3⌟ A round mound on the north of the (1).

   (3) ⌞1.5;0.2⌟ Very small peak.

   (4) ⌞6/1.5;<0.2⌟ (5) ⌞5̇/2;0.3⌟ Uncinate hills, which are probably walls of an old crater.

   (6) ⌞4;0.3⌟ A rather distinct mound.

   (7) ⌞10/3;<0.3e⌟ (8) ⌞10/3;<0.1e⌟ Linearments parallel to each other.

   (9) ⌞4;0.6⌟ A distinct peak on the bottom of the north wall.

   (10) ⌞4;<0.3e⌟ A round mound on the north wall.

   (11) ⌞4;0.4⌟ A round mound on the bottom of the north-west wall.

   (12) ⌞3;<0.2⌟ A small mound.

   (13) ⌞6/3;<0.2⌟ (14) ⌞10/2;<0.2⌟ Concentric ridges around the (11).

   There are grooves of length 10 km on the east of the (7), and on the west of the (8). There is a depression between the (10) and (14).

13. Franklin (+576,+627;56)

   (1):γ ⌞5̇;1.0⌟ A typical central peak.

   (2) ⌞2;U. L.⌟ Small mound on bottom of the (1).

Franklin

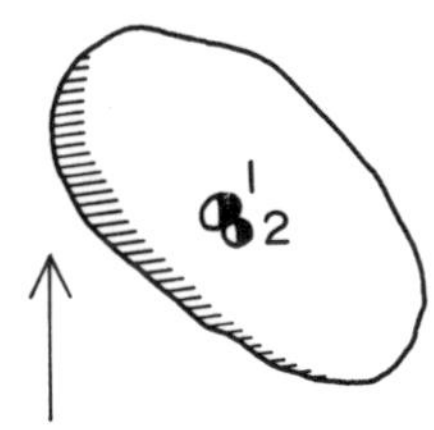

14. Galle (+213,+827;21)

   At the center of the floor, there is a small, but deep pit whose diameter is about 1 km.

   (1) ⌞5̇/3;U. L.⌟ An oblong elevation.

   (2) ⌞8;0.5w⌟ Large elevated part on the north wall.

   (3) ⌞1;U. L.⌟ A mound on the south-west side of the floor pit mentioned above.

Galle

(Based on the Lunar Orbiter IV Photographs)

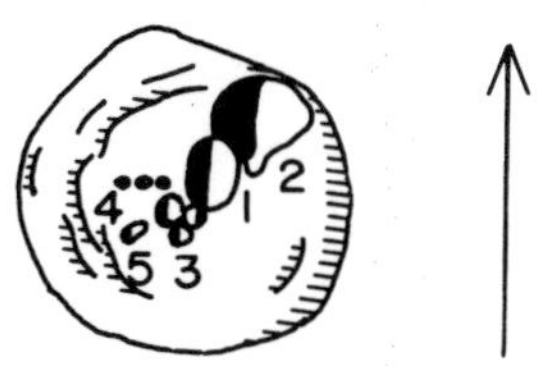

   (4) ⌞3;U. L.⌟ A low hill.

   (5) ⌞2;U. L.⌟ A faint patch.

15. Grove (+415,+647;27)

   (1) ⌞2;U. L.⌟ A white patch on the center of the floor.

Hayn

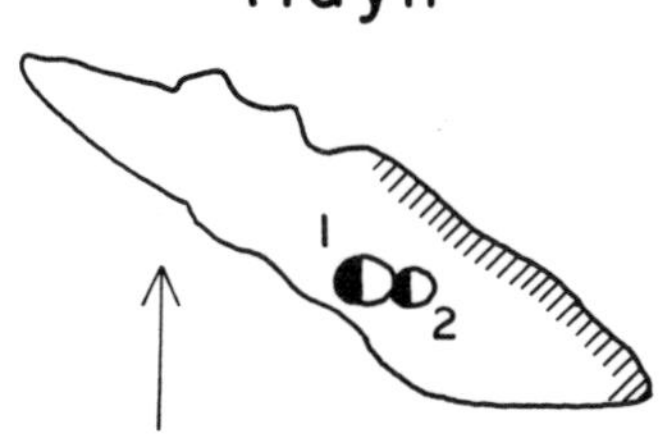

16. Hayn (+426,+904;87)

   (1) ⌞15;2⌟ A central peak having conical form.

   (2) ⌞10;2⌟ A large round mountain on the east of the (1).

17. Hayn A (+430,+890;54)

   (1) ⌞12;1⌟ A central mountain having easterly

steep slope.

18. Hayn E (+357,+921;41)

(1) ⌜5̇;2⌝ A central peak having conical form.

19. Hercules (+434,+728;67)

The floor is very smooth.

(1) ⌜2.5;0.3⌝ A tiny peak. On the west bottom of it, there is a white patch.

20. Hercules A (+432,+779;33)

White patches are seen on the bottom of the south-west wall. They may be also faint elevations.

(1) ⌜5;U. L.⌝ A very faint elevation on the bottom of the west wall.

Mitchell

21. Mitchell (+223,+763;30)

The wall of this formation and that of Aristoteles are mutually penetrated.

(1) ⌜4.5;0.4⌝ (2) ⌜3;0.2⌝ (3) ⌜3;0.4w⌝ The (1) and (3) are faint peaks, which form respectively west and east wall of a hill-top crater, whose diameter is about 5 km. The (2) is a hill, which is a collapsed part of the east wall.

(4) ⌜4/3;0.5e⌝ A low hill.

(5) ⌜3;0.3⌝ A low mound on the bottom of the south wall.

22. Plana (+350,+672;44)

(1) ⌜8;0.7⌝ A very distinct central peak. In its east bottom, very small hill is attached.

23. Schwabe F, U. F. on the N. of (+317,+920;48)

(1) ⌜5;0.6⌝ A distinct central peak.

1. <u>Anaxagoras</u> (-050,+959;51)
   (1) ⌞10;1⌟ A large mountain, having three peaks, stands on the southerly eccentric position of the floor.
2. <u>Archytas</u> (+045,+855;32)
   The floor has faint elevations and is a little rugged.
   (1) ⌞4;0.6⌟ A peak on the south-easterly eccentric position of the floor.
3. <u>Archytas G</u> (+005,+825;7)
   (1) ⌞5;U. L.⌟ Very minute concentric crater.
   The south-adjacent crater is also likely to be a concentric crater, according to the Photographs of the Lunar Orbiter IV.

Aristoteles

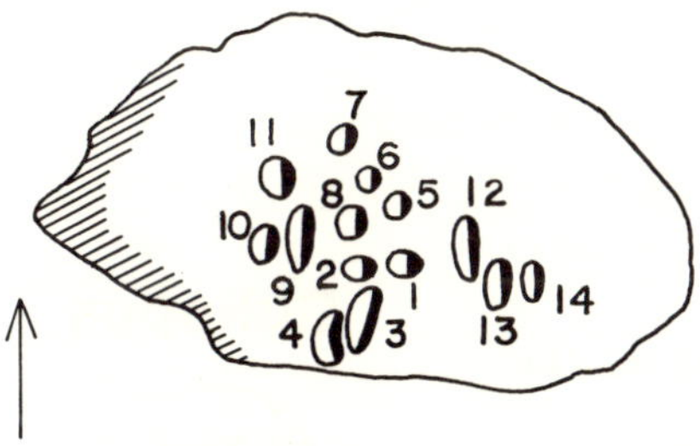

4. <u>Aristoteles</u> (+191,+768;87)
   (1):θ ⌞6;0.$\dot{5}$⌟ The most distinct peak, standing on the southerly eccentric position of the floor.
   (2):ζ ⌞3;0.3⌟ A distinct ridge standing to the west of the (1).
   (3) ⌞17/3.5;U. L.⌟ A faint hill connecting the peak (1) with the south wall.
   (4) ⌞11/3;0.6w⌟ A distinct uncinate dyke on the south wall.
   (5) ⌞3;U. L.⌟ A hill placed nearly on the center of the floor.
   (6) ⌞2;U. L.⌟ A faint mound.
   (7) ⌞3;U. L.⌟ Very minute mountain near the north wall.
   (8) ⌞2;U. L.⌟ A faint patch.
   (9) ⌞10/3;U. L.⌟ (10) ⌞10/3;U. L.⌟ Very faint ridges.
   (11) ⌞3;U. L.⌟ A faint elevation.
   (12) ⌞10/4;U. L.⌟ (13) ⌞11/4;U. L.⌟ (14) ⌞8/3;U. L.⌟ Faint parallel ridges.
   These descriptions are confirmed by the photographs of the Lunar Orbiter IV.
5. <u>Birmingham</u> (-078,+905;97)
   The floor is very rugged, and only three peaks are remarkable.
   (1) ⌞$\dot{5}$;0.3⌟ A peak standing on the center of the floor.
   (2) ⌞4;0.2⌟ A peak standing on the south of a small crater, Birmingham G.
   (3) ⌞8/4;0.2⌟ An oblong hill at the south-westerly part of the floor.

Birmingham

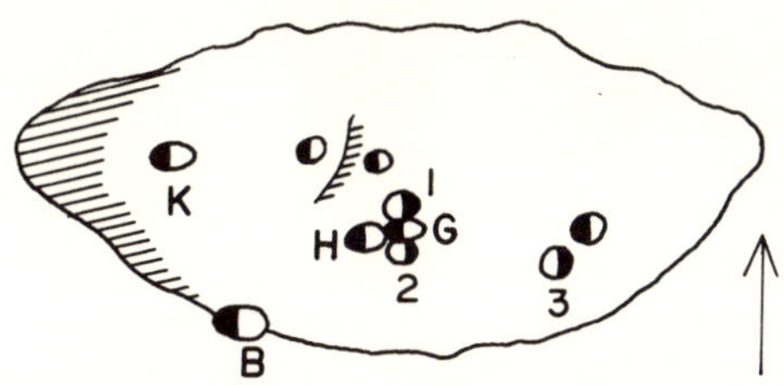

6. <u>Bond, W. C.</u> (+026,+909;157)
   The floor is very rich in small mounds. Here, only hills near the center of the floor are registered.
   (1) ⌞3.5;0.2⌟ Conical peak. Probably, the true central peak. It is whitish.
   (2) ⌞3;0.2⌟ (3) ⌞4;0.2⌟ (4) ⌞4;0.2⌟ (5) ⌞4;0.2⌟ (6) ⌞4;0.2⌟ Low mounds standing in an arc.

(XV)

(7) ⌊5̇;0.1⌋ A low hill having a summit crater.

(8) ⌊2.5;0.1⌋ A round mound.

(9) ⌊2.5;0.1⌋ (10) ⌊2.5;0.1⌋

(11) ⌊4;0.1⌋ Low hills forming a triangle.

(12) ⌊3;0.1⌋ (13) ⌊2.5;0.1⌋

(14) ⌊3;0.1⌋ (15) ⌊3;0.1⌋ Small mounds standing like the "Trapesium".

(16) ⌊2;0.1⌋ A small mound.

(17) ⌊30/5;0.1⌋ A long hill, southerly broader.

(18) ⌊6;0.1⌋ (19) ⌊4;0.1⌋ Broad hills.

(20) ⌊4;0.1⌋ A small hill.

Bond, W.C.

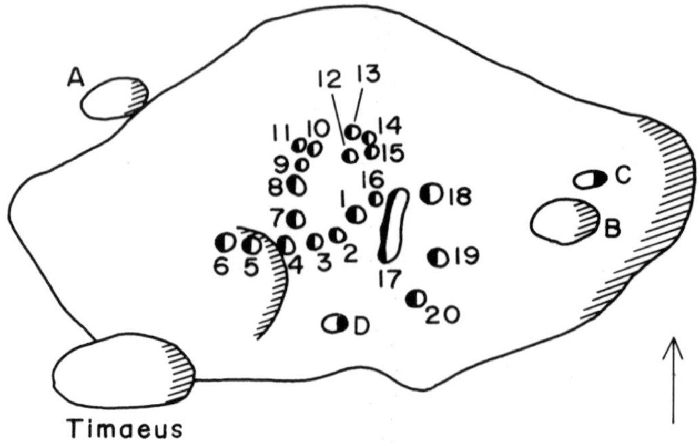

7. <u>Calippus</u> (+145,+629;31)

There are several faint mounds on the floor.

(1) ⌊4.0;<0.2⌋ Faint and whitish mound.

(2) ⌊4.5;<0.4⌋ Rather distinct mound at the bottom of the north-west wall.

(3) ⌊4.5;<0.2⌋ A faint mound at the bottom of the north-west wall.

(4) ⌊5̇;<0.2⌋ Faintly elevated part of the floor.

(5) ⌊5̇;<0.4⌋ (6) ⌊4;<0.2⌋ (7) ⌊4;<0.2⌋ A row of mounds standing along the south-west wall.

Calippus

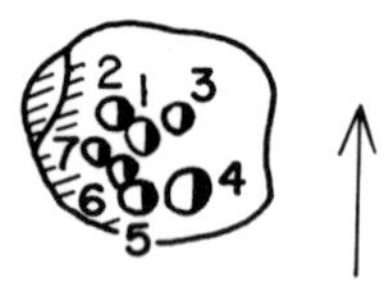

8. <u>Calippus A</u> (+110,+602;16)

(1) ⌊3;0.3⌋ A very tiny peak on the south-westerly eccentric position.

Cassini, A

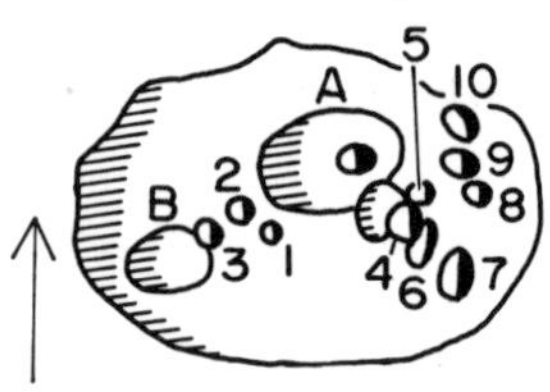

9. <u>Cassini</u> (+061,+646;57)

(1) ⌊4;0.1⌋ A very low peak.

(2) ⌊4;0.1⌋ (3) ⌊3;0.1⌋ Low mounds.

(4) ⌊8/1.5;0.6e⌋ A hill which is a part of wall of an unnamed crater on the south of Cassini A, whose diameter is about 7 km.

(5) ⌊7/1.5;0.3⌋ Small low hill on the north bottom of the (4).

(6) ⌊7/2.5;0.1e⌋ A narrow ridge.

(7) ⌊3.5;0.3⌋ The highest part of a mountain block on the south-east of the floor.

(8) ⌊4;0.4⌋ A peak on the bottom of the east wall.

(9) ⌊5̇;0.3⌋ Low mountain.

(10) ⌊4;0.4⌋ A distinct peak. It is likely to be an end of the north wall.

10. <u>Cassini A</u> (+063,+649;17)

A compound crater.

(1) ⌊5;U. L.⌋ An elevated part of the floor at easterly eccentric position

of the floor, and may be trace of west wall of smaller component.

11. De Sitter (+110,+984;65)

This formation was called formerly Euctemon J.

(1) ⌞18/5;0.4⌟ A long dyke runs through the floor in north-south direction, on whose central part, stands a central peak.

12. Epigenes (-032,+923;55)

The west part of the floor is flat.

(1) ⌞8;0.45⌟ A large mountain block, composed by two components.

(2) ⌞8;0.4⌟ Another mountain on the south of the (1).

(3) ⌞2.5;0.05⌟ A very small peak.

(4) ⌞4;<0.2⌟ A low mound.

(5) ⌞6;<0.2⌟ A low hill.

(6) ⌞4;0.2⌟ A peak on the bottom of the east wall.

(7) ⌞4;0.2⌟ A hill connected with the (6).

Epigenes

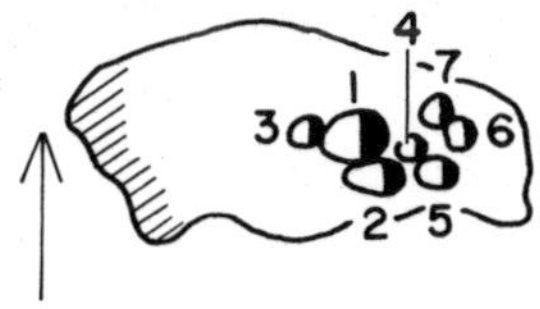

13. Fontenelle (-145,+893;38)

(1) ⌞9;0.3⌟ A small crater with elevated wall, on slightly easterly eccentric position of the floor.

(2) ⌞2.5;U. L.⌟ (3) ⌞4;U. L.⌟ (4) ⌞4;U. L.⌟ Very faint mounds standing on a line.

(5) ⌞4;U. L.⌟ Rather distinct hill.

(6) ⌞3;U. L.⌟ Small faint mound.

(7) ⌞3.5;U. L.⌟ Higher part of elevated floor, and may be a trace of terrace structure of the east wall.

Fontenelle

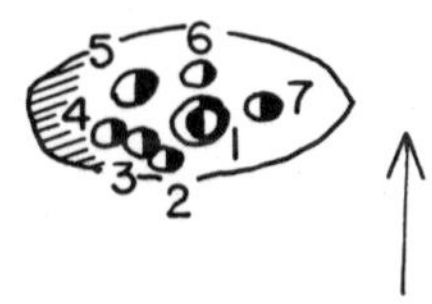

14. Goldschmidt (-015,+957;125)

There are innumerable mounds on the floor.

(1) ⌞6.5;0.3⌟ (2) ⌞5̇;0.3⌟ These mounds form east and west wall of a small crater, respectively.

(3) ⌞2.5;0.1⌟ (4) ⌞2;0.1⌟ (5) ⌞3;0.2⌟ Mounds which are parts of wall of a small crater.

Goldschmidt

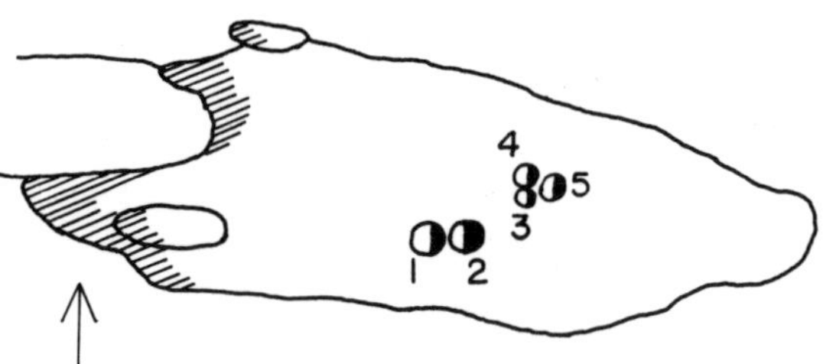

15. Mayer, C. (+134, +893;38)

(1) ⌞16/6;0.6⌟ Two-peaked ridge.

(2) ⌞5̇;U. L.⌟ Round mound on the north wall.

C. Mayer

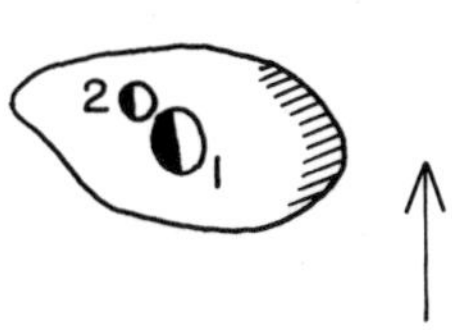

16. Mayer, C. B (+132,+867;36)

(1) ⌞4.5;0.1⌟ Very low central mound.

17. Mouchez (-092,+979;82)

The floor is very rugged.

(1) ⌞4;0.5̇⌟ A peak on the westerly eccentric position of the floor.

Mouchez

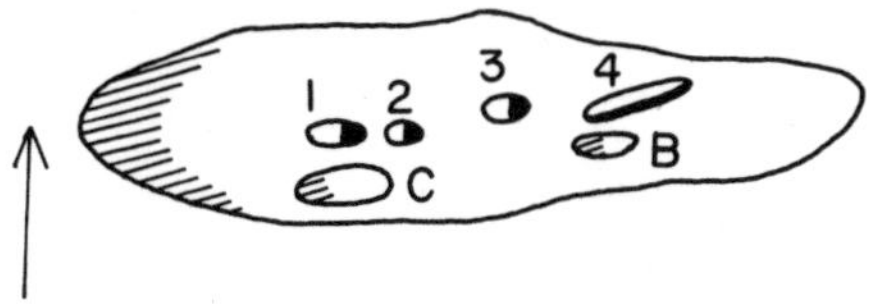

(XV)

(2) ⌞3.5;0.3⌟ A small peak, which may be the true central peak.

(3) ⌞4;0.3⌟ A low hill.

(4) ⌞25/7;0.3⌟ A long dyke extended towards the north wall of Mouchez B.

Mouchez A

18. Mouchez A (-080,+987;51)

(1) ⌞6/3;<0.5⌟ An oblong hill.

(2) ⌞3;<0.5⌟ A round mound.

(3) ⌞4;<0.5e⌟ Collapsed part of the east wall.

19. Philolaus (-165,+951;71)

(1) ⌞12;1.2e⌟ Large mountain.

(2) ⌞1̇0;0.7e⌟ (3) ⌞5̇;0.5̇e⌟ Peaks standing on the north side of the (1).

(4) ⌞8;1.3e⌟ The highest peak standing on the easterly eccentric position.

(5) ⌞14/4;0.2⌟ A long dyke.

(6) ⌞6;0.4⌟ A low hill at the bottom of the east wall.

Philolaus

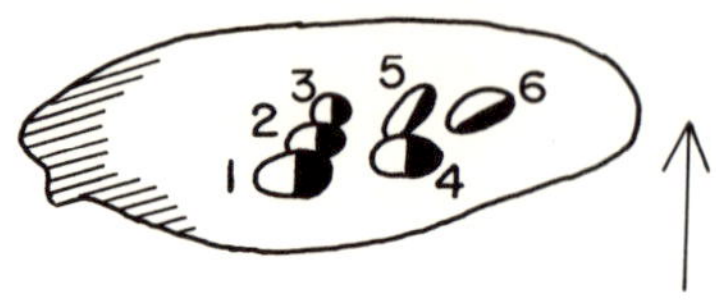

U.F. on the South-East of Philolaus D

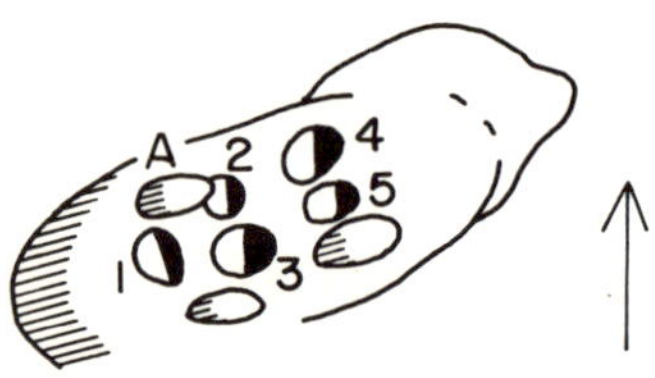

20. Philolaus D, U. F. on the S.-E. of (-104,+954;49)

There are innumerable hills on the floor. Here, only distinct mountains are registered.

(1) ⌞8;0.4⌟ A peaked mountain.

(2) ⌞1̇0;0.8⌟ A mountain forming the east wall of Philolaus A.

(3) ⌞8;0.6⌟ Low mountain.

(4) ⌞6;0.4⌟ Probably collapsed part of the wall.

(5) ⌞6;0.3⌟ Another peaked mountain.

21. Protagoras (+071,+828;22)

This formation has a terrace structure on inner side of the wall and looks like a double crater.

(1) ⌞2.5;U. L.⌟ A faint mound on the center of the floor.

22. Scoresby (+052,+978;56)

(1) ⌞6;1.3⌟ Probably the true central peak.

(2) ⌞6;1.3⌟ Neighboring peak of the (1).

(3) ⌞5;0.5e⌟ A hill on the slope of the north wall.

Scoresby

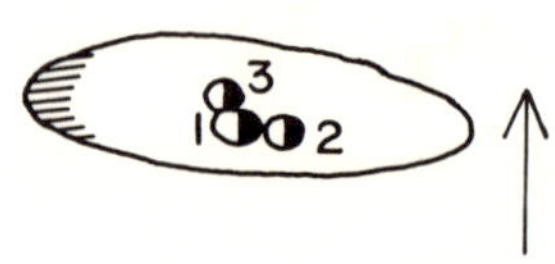

23. Sheepshanks (+150,+859;25)

(1) ⌞4;U. L.⌟ A faintly elevated part on the slope of the north wall.

24. Theatetus (+084,+602;25)

(1) ⌞5;U. L.⌟ Very faintly elevated part on the north-west part of the floor.

(2) ⌞1.5;U. L.⌟ Very faint mound, but may be the true central peak.

(3) ⌞2.5;U. L.⌟ A faint and oblong hill.

25. Timaeus (-004,+890;33)

(XV)

Theaetetus

(1) ⌞3;0.8⌟ A peak.

(2) ⌞3;U. L.⌟ Very faint elevation.

Timaeus

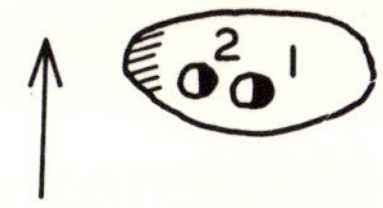

XVI

1. Bianchini (-372,+752;39)
   (1) ⌞2.5;0.3⌟ Very tiny peak on rugged floor.
2. Carpenter (-273,+936;60)
   This formation has elevated circumference like that of Tycho.
   (1) ⌞10;1.0⌟ A large mountain block, on which several peaks are perceived.
   (2) ⌞5;0.5⌟ According to photographs of the Lunar Orbiter IV, there is another small mountain behind the (1), which is not visible from the earth.

Carpenter

3. Harpalus (-416,+795;40)
   (1) ⌞2.5;0.4⌟ (2) ⌞2.5;0.3⌟ Two mounds standing side by side.
   (3) ⌞2;0.2⌟ A smaller hill to the east of the (1) and (2).

Harpalus

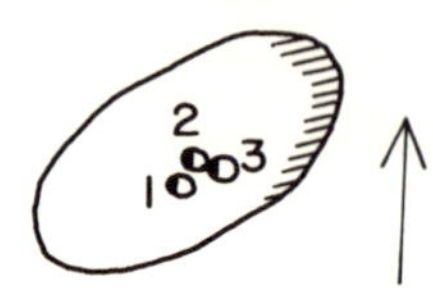

4. Mairan (-514,+664;41)
   (1) ⌞3;0.3⌟ (2) ⌞2;0.2⌟ These two peaks are standing in a north-south line. The north component, (2), is lower, but stands on the center of the floor. Probably it is the true central peak.
   (3) ⌞2;0.2⌟ Another small hill to the north of the (2), according to the photographs of the Lunar Orbiter IV.

Mairan

5. Markov (-530,+802;41)
   (1) ⌞13;1⌟ A very large, but gentle-formed, mountain block.
6. Pascal (-251,+963;102)
   (1) ⌞1̇0/3;0.2⌟ An oblong hill along the north-south direction.
7. Pythagoras (-397,+894;128)
   (1) ⌞20;3.6⌟ (2) ⌞20;1.8⌟ (3) ⌞4;0.1e⌟
   :*α* The (1) is large, peaked mountain. On the north of it, another large mountain (2) is attached. The mound (3) is seen on the west slope of the (1). These three objects are generally regarded as one mountain block, and named *α*.

Pythagoras

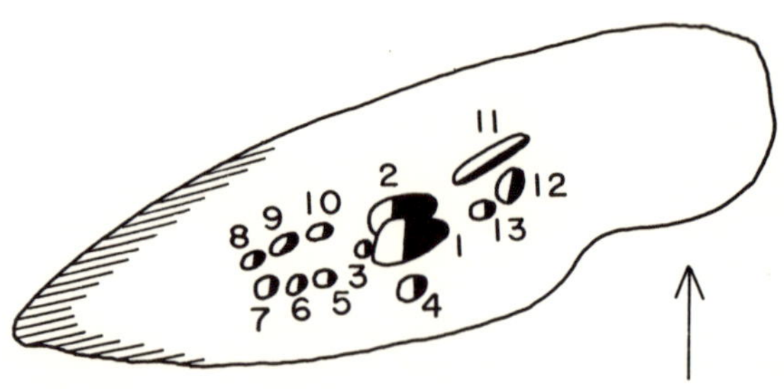

   (4) ⌞5̇;U. L.⌟ A faint elevation to the south of the (1).
   (5) ⌞4;0.2⌟ (6) ⌞4;0.3⌟ (7) ⌞3;U. L.⌟ Faint hills, standing in a line.
   (8) ⌞3;U. L.⌟ (9) ⌞5̇;U. L.⌟ (10) ⌞3;U. L.⌟ Segments of a faint scarp.
   (11) ⌞11/2;U. L.⌟ A low dyke.
   (12) ⌞5̇;U. L.⌟ A faint elevation.
   (13) ⌞3;0.2⌟ Tiny hill on the east of the (1).
8. Pythagoras, U. F. to the W. of (-430,+888;26)

(XVI)

(1) ⌞3;0.2⌟ A low mound at the center of a very indistinct formation.

9. Sharp (-451,+716;40)

(1) ⌞3;U. L.⌟ (2) ⌞3;U. L.⌟ There is a swarm of small pits on the center of the floor. The hill (1) and (2) are disturbed features by these pits.

Sharp

(Based on the Lunar Orbiter IV Photographs)

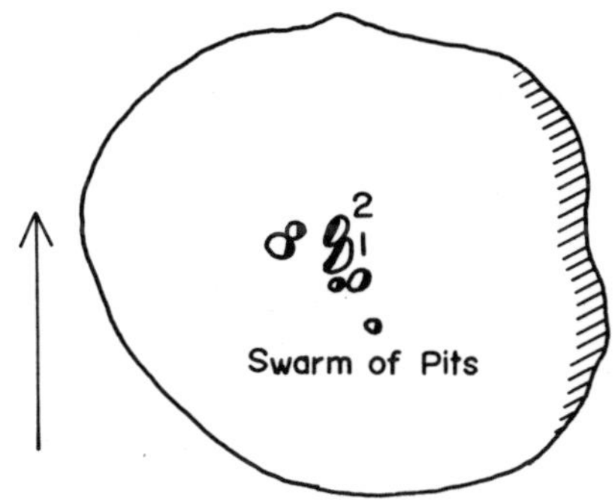

10. Sharp B (-485,+731;22)

(1) ⌞8/3;0.3⌟ (2) ⌞8/3;0.3⌟ These two ridges are almost parallel.

(3) ⌞5̇/3;0.1⌟ A ridge standing on the north of the (2). These ridges are parts of rugged feature on the north part of the floor.

11. Volta (-585,+809;109)

(1) ⌞4;U. L.⌟ A very low central mound.

12. Xenophanes (-532,+842;111)

(1) ⌞10̇;1.1⌟

(2) ⌞7;1.4⌟

(3) ⌞6;1.2⌟ Three peaks on a long ridge.

(4) ⌞6;U. L.⌟ Another hill on an extension of the ridge (3).

(5) ⌞4;1.4⌟ A lofty peak.

Xenophanes

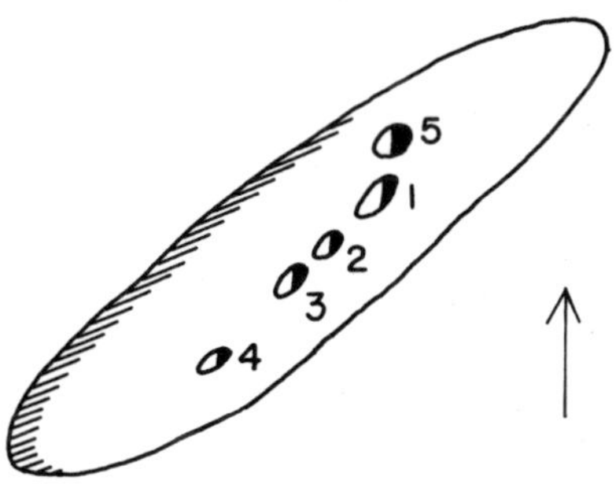

Sharp B

(Based on the Lunar Orbiter IV Photographs)

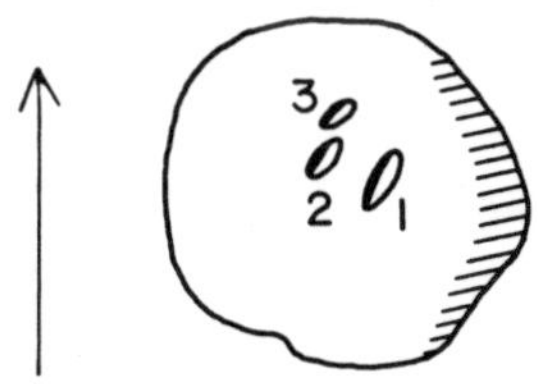

XVII

1. Galvani (-644,+762;75)
   (1) ⌊4;U. L.⌋ A faint peak on the northerly eccentric position of the floor. The floor is rich in cracks, according to the Lunar Orbiter IV Photographs.
2. Gerard Q (-686,+721;177)
   (1) ⌊61;U. L.⌋ A large concentric formation.
3. Lavoisier A (-764,+600;27)
   (1) ⌊3;<0.7⌋ A low mound on the easterly eccentric position.
4. Lavoisier D (-737,+656;62)
   (1) ⌊15/9;<0.5⌋ (2) ⌊10;U. L.⌋ A long faint ridge runs in north-south direction, of which northern part is the (1), and the southern part is the (2). The former one is somewhat higher, and connected to the north wall.
   (3) ⌊6;<1⌋ A small peak at the bottom of the south-east wall.

Lavoisier D, E

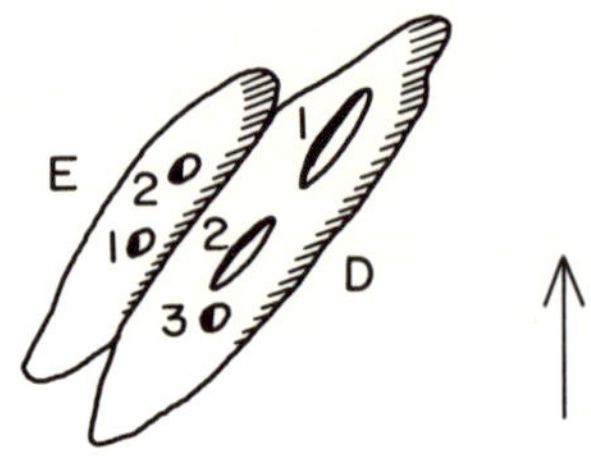

5. Lavoisier E (-745,+653;48)
   (1) ⌊4.5;0.7⌋ A low central peak.
   (2) ⌊4.5;0.3⌋ A hill at the bottom of the north-west wall.
6. Repsold (-612,+779;107)
   (1) ⌊20/5;1.1⌋ An oblong mountain who has at least two peaks.
7. Repsold G (-626,+771;44)
   (1) ⌊11;U. L.⌋ A low, but large central mountain.
8. Repsold T (-663,+739;12)
   (1) ⌊3;U. L.⌋ A very faint central mountain.
9. Repsold T, U. F. to the S.-E. of (-666,+733;19)
   (1) ⌊4;0.5⌋ A faint hill on the southerly eccentric position of the floor.

1. Aristarchus (-676,+402;45)
   (1) ⌞3;<0.2⌟ A very low central mound.
2. Balboa A (-944,+299;47)
   (1) ⌞3.5;1⌟ A typical central peak.
3. Briggs (-835,+445;39)
   (1) ⌞13/4;0.2⌟ (2) ⌞5̇;0.1⌟:β The (1) is a peaked hill. The (2) is a broad low mountain, standing at the center of the floor. These two objects are generally regarded as one mountain block and named as β .
   (3) ⌞6;0.3⌟ A broad elevation on the eastern part of the floor.
4. Briggs B (-832,+471;25)
   (1) ⌞3;0.3⌟ A small peak disturbed by a scarp, according to the photographs of the Lunar Orbiter IV.

Briggs

(Based on the Lunar Orbiter IV Photographs)

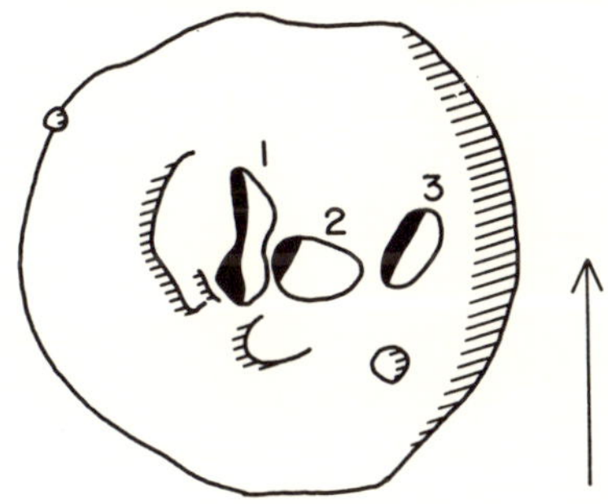

Briggs B

(Based on the Lunar Orbiter IV Photographs)

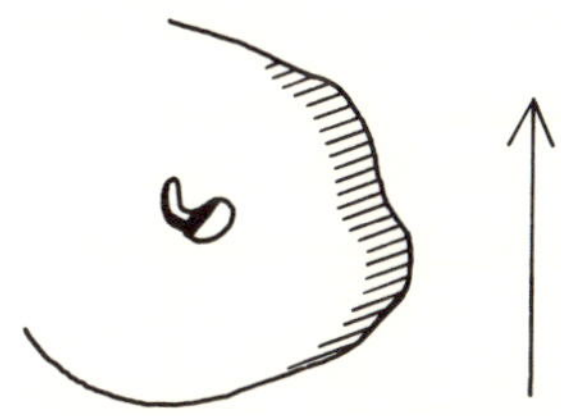

Lichtenberg, U.F. to the N. of Lichtenberg

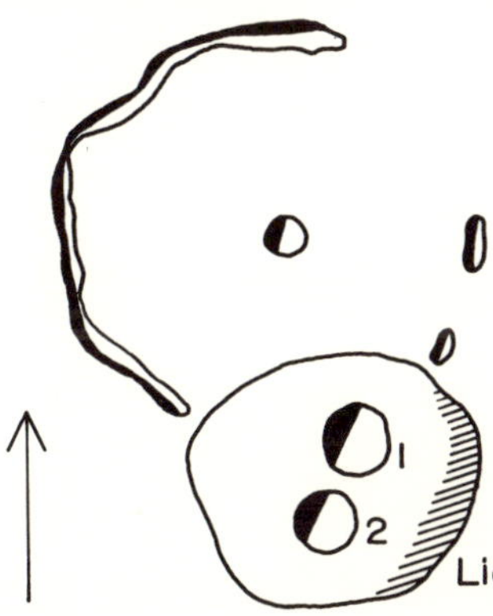

5. Lichitenberg (-785,+527;21)
   This formation may be a compound crater.
   (1) ⌞4;U. L.⌟ (2) ⌞5̇;U. L.⌟ Both (1) and (2) are faint elevations.
6. Lichtenberg, U. F. on the N. of (-780,+537;32)
   This formation is a mountain ring superposed by Lichtenberg.
   (1) ⌞3;0.3⌟ A round mound.
7. Marius (-758,+206;41)
   (1) ⌞2;U. L.⌟ A white patch is seen on somewhat southerly eccentric position.
8. Röntgen (-839,+545;97)
   (1) ⌞4;U. L.⌟ A small central peak.
   (2) ⌞1;U. L.⌟ A peak on north-east side of the (1).
9. Schiaparelli (-784,+396;24)
   (1) ⌞5;U. L.⌟ Very faint elevation on the westerly eccentric position.
10. Seleucus (-856,+360;43)

(XVIII)

(1) ⌐4;U. L.⌐ A white patch on the floor.

Röntgen

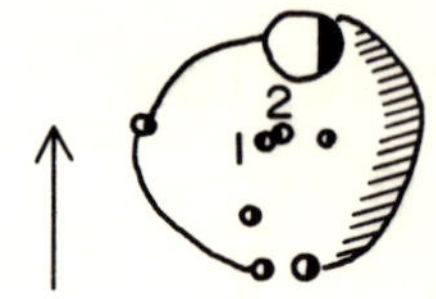

XIX

1. Cavalerius (-916,+089;64)

(1) ⌈3;1.1⌋ The main hill standing on the southerly eccentric position.

(2) ⌈3;U. L.⌋ To the north of the (1), stands very faint hill (2), which may be the true central peak.

Cavalerius

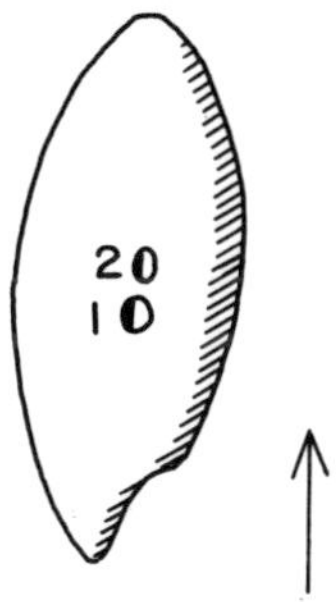

2. Damoiseau (-872,-086;37)

(1) ⌈3;<1⌋ A very low peak.

(2) ⌈3;U. L.⌋ A faint peak on the north part of the floor.

(3) ⌈4;U.L.⌋ A white patch.

Damoiseau

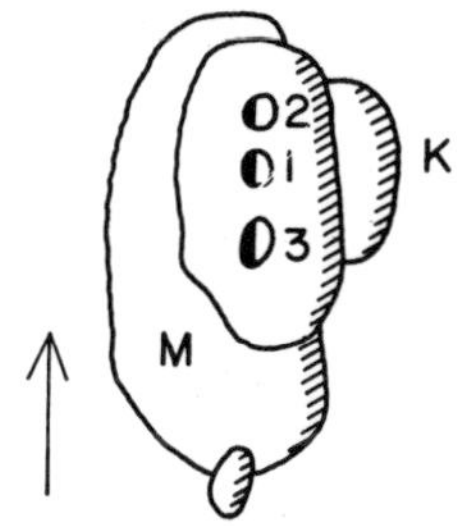

3. Flamsteed (-696,-078;21)

(1) ⌈4;U.L.⌋ Aggregation of very minute hills, according to the Lunar Orbiter IV Photographs.

Flamsteed (Based on the Lunar Orbiter IV Photographs)

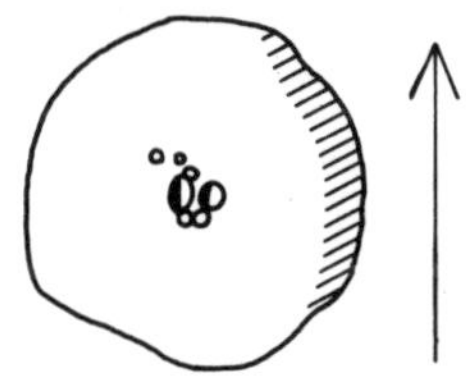

4. Grimaldi (-925,-091;222)

(1) ⌈3;U. L.⌋ The most distinct mound on the northern part of the floor.

(2) ⌈1;U. L.⌋ (3) ⌈3;U. L.⌋

(4) ⌈2;U. L.⌋ (5) ⌈2;U. L.⌋

(6) ⌈5;U. L.⌋ (7) ⌈1;U. L.⌋

The minute hills from (2) to (6) are remnant of collapsed crater, whose

Grimaldi (Based on the Lunar Orbiter IV Photographs)

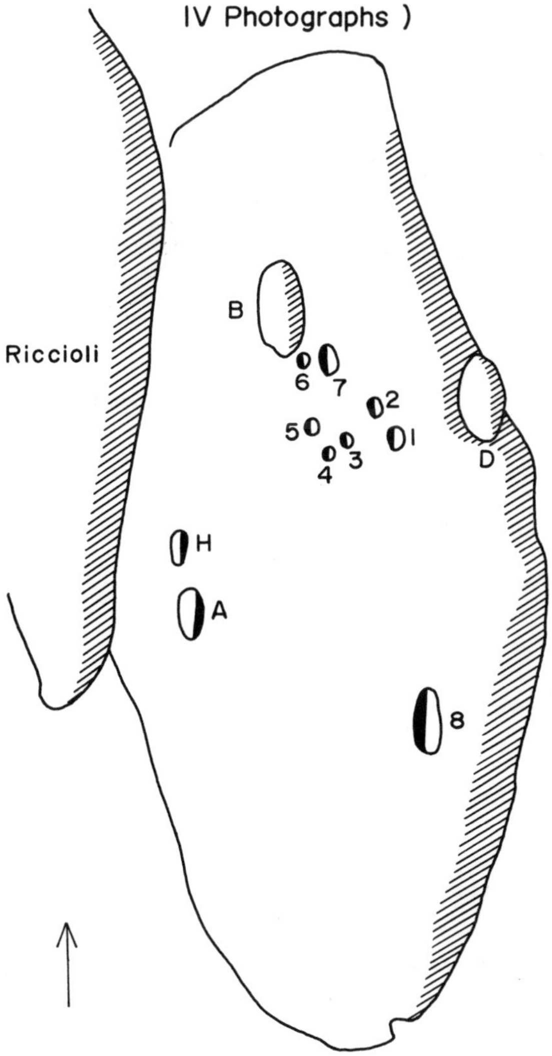

Hansteen

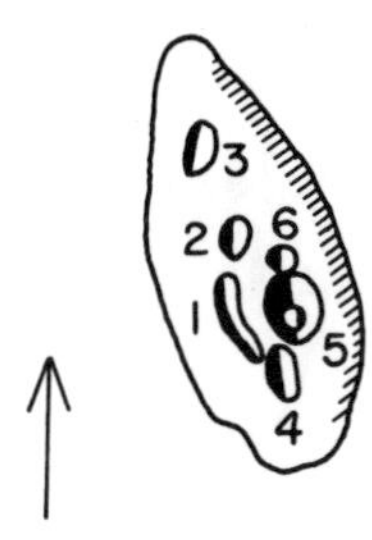

central peak is the mound (7).

(8) ⌊28/2;U. L.⌋ A ridge having westerly steep slope.

5. Hansteen (-771,-200;45)

(1) ⌊10/5;0.2⌋ A long ridge.

(2) ⌊2;0.1⌋ A low peak.

(3) ⌊5̇/3;0.2w⌋ A collapsed part of the north wall.

(4) ⌊5̇/3;U. L.⌋ Another ridge.

(5) ⌊7;U. L.⌋ A low mountain with a hill-top crater.

(6) ⌊3.5;U. L.⌋ A faint round mound.

6. Hevelius (-923,+038;118)

(1):α ⌊8/6;1.0⌋ Very low hill, which is oblong to the north-east and south-west direction.

Kepler

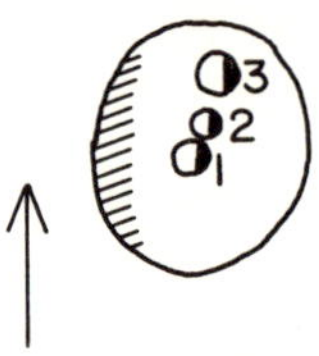

7. Kepler (-609,+141;32)

(1) ⌊4;U. L.⌋ A low mound, probably the true central peak.

(2) ⌊3.5;0.2⌋ Another mound on the northerly eccentric position of the floor.

(3) ⌊4;0.3⌋ The largest mound on the north wall.

Letronne

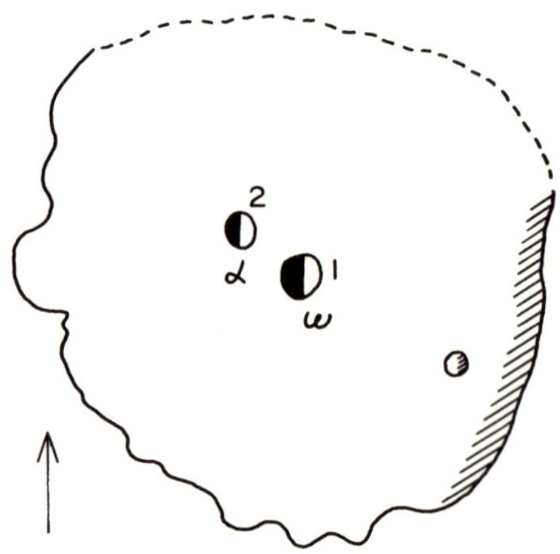

8. Letronne (-662,-184;119)

(1):ω ⌊10;0.2⌋ Very low, peaked mountain on a faint ridge.

(2):α ⌊5;U. L.⌋ White patch, whose height is not perceivable. It may be a remnant of small crater.

9. Lohrmann (-923,-008;34)

(1) ⌊8;U. L.⌋ Very faint elevation of the west part of the floor.

Olbers

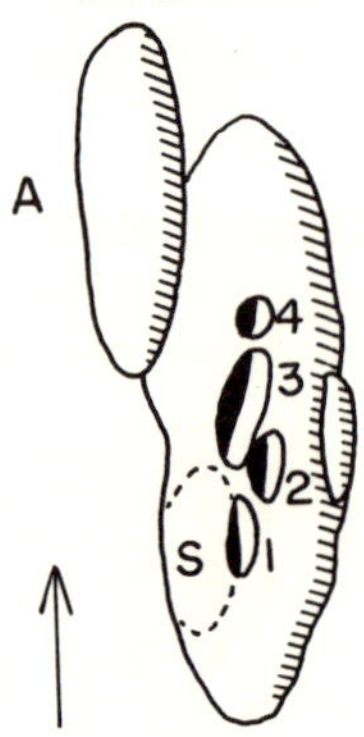

10. Olbers (-962,+124;71)

(1) ⌊10;0.8⌋ A ridge, which may be a part of the crater Olbers S.

(2) ⌊10;U. L.⌋ A faint ridge.

(3) ⌊12;U. L.⌋ Very faint elevation.

(4) ⌊4; U. L.⌋ A low hill.

11. Olbers, U. F. to the S. of (-972,+054;143)

Formerly, this formation was called Hedin.

(1) ⌊25/20;1⌋ A long ridge.

(2) ⌊20/7;U. L.⌋ A low hill.

(3) ⌊10;U. L.⌋ (4) ⌊10;U. L.⌋ Faint elevations.

(5) ⌊10;1⌋ A mountain neighboring a shallow groove.

(6) ⌊15;1⌋ (7) ⌊20;1⌋ Extensions of the south wall.

(8) ⌊25/7;U. L.⌋ A long ridge.

(9) ⌊5̇;1⌋ Rather distinct peak.

(10) ⌞10;1⌟ A distinct mountain at the bottom of the north wall.
According to the author's opinion, the west wall of this formation and the ridge (1), (5), (7) and (10), form an independent crater, and the elevation (4) is its central peak.

U.F. to the S.of Olbers

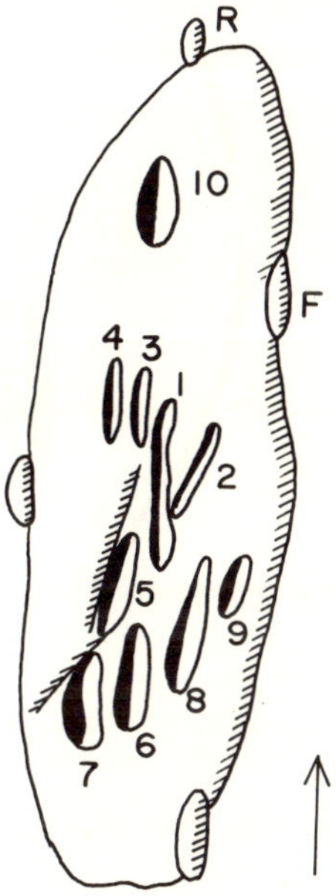

12. Olbers, U. F. to the S.-W. of (-974,+095;63)
This formation was formerly called Hedin A.
(1) ⌞20;U. L.⌟ A very broad and low mountain covers the southern part of the floor.

13. Olbers D, U. F. on the W. of (-968,+172;34)
(1) ⌞8;U. L.⌟ A faint hill, connected with a dyke which lies on the north wall.

14. Reiner (-812,+120;30)
(1) ⌞4;0.2⌟ (2) ⌞7;0.2⌟:α
The main peak (1) stands on somewhat northerly eccentric position. The low and broad mountain (2) is situated on the south of (1). These two objects are ordinarily referred as α, but Wilkins and Moore described in their book "The Moon", that the peak α splits into two hills.
(3) ⌞8;U. L.⌟ A broad elevation on the inner slope of the south wall.

Reiner
(Based on the Lunar Orbiter IV Photographs)

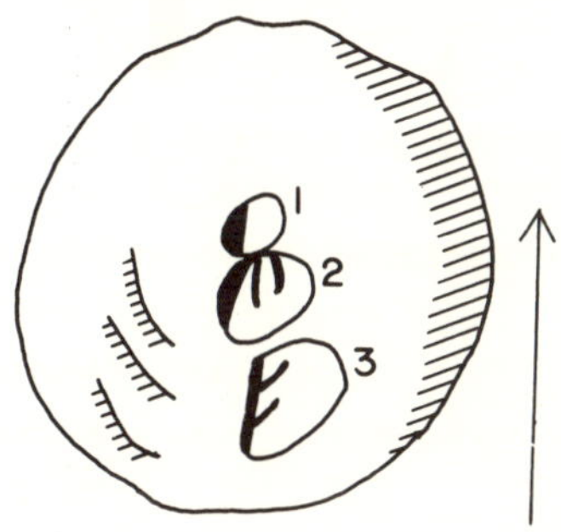

15. Reiner G (-814,+051;15)
At the center of the floor, there is a small pit whose diameter is about 0.8 km.
(1) ⌞4/1;0.1⌟ A dyke, which may be remnant of wall of a floor crater.
(2) ⌞0.5;U. L.⌟ (3) ⌞0.5;U. L.⌟ On the north and west of the central pit, there are small mounds (2) and (3) respectively.
(4) ⌞3;0.3⌟ A broad hill.

Reiner G
(Based on the Lunar Orbiter IV Photographs)

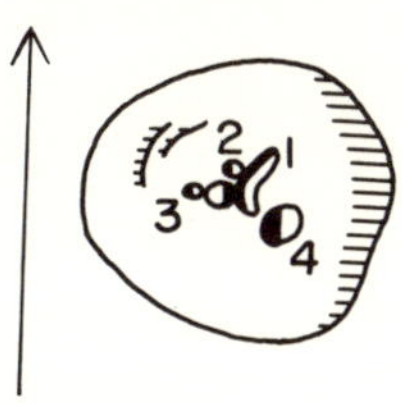

16. Riccioli (-961,-054;152)
The north part of the floor is dark and smooth, which is named Riccioli D. The other part of the floor is very rugged.
(1):δ ⌞20/9;0.8⌟ A ridge-like central mountain.
(2):ρ ⌞12;0.6⌟ Secondly eminent hill.
(3) ⌞4;0.6⌟ (4) ⌞4;U. L.⌟ Low hills.
(5) ⌞16;0.8w⌟ High ridge, which may be west wall of riccioli S.
(6) ⌞5̇;U. L.⌟ A low hill.
(7) ⌞12;1.2⌟ A high ridge.

(XIX)

Riccioli

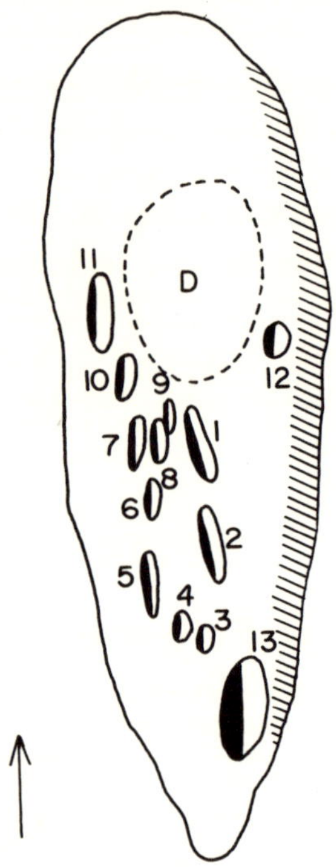

(8) ⌞8;0.6⌟ (9) ⌞7;0.5⌟ Hills.

(10) ⌞10;1.4⌟ Broad elevation.

(11) ⌞16/10;0.4w⌟ A high ridge, which may be a collapsed part of the west wall.

(12) ⌞5;0.4⌟ An eminent peak.

(13) ⌞22;1.6⌟ A large and dark mountain block.

17\. <u>Schlüter</u> (-988,-105;91)

(1) ⌞10;3⌟ Very large central peak.

Sirsalis E

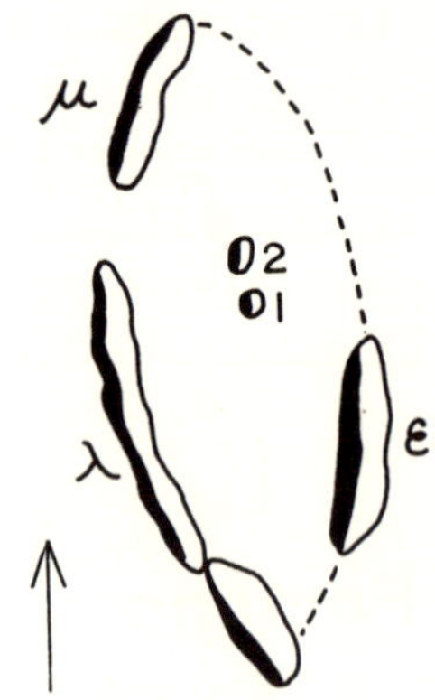

18\. <u>Schlüter A</u> (-978,-160;35)

(1) ⌞10;U. L.⌟ Very faint central elevation.

19\. <u>Sirsalis E</u> (-825,-142;72)

(1) ⌞2.8;<0.1⌟ Faint hill, which may be the true central peak.

(2) ⌞4.4;<0.1⌟ Faint, but larger hill standing on the northerly eccentric position of the floor.

20\. <u>Vasco da Gama R</u> (-978,+172;56)

(1) ⌞26/10;U. L.⌟ A long and faint ridge runs in north-south direction. Its southern parts are higher.

1. Bouvard A (-806,-586;68)
   (1) ⌞6;1⌟ A small hill on the bottom of the north wall.
2. Darwin B (-895,-341;51/39)
   (1) ⌞8;0.4⌟ A low hill.
   (2) ⌞16/10;0.4⌟ A ridge which is collapsed part of the north-west wall.
3. Doppelmayer G, U. F. on the S. of (-618,-492;17)
   (1) ⌞4;U. L.⌟ Very faint central mound.
4. Eichstadt (-904,-385;49)
   (1) ⌞7;U. L.⌟ Very low central peak.

Darwin B

5. Fontana (-802,-277;31)
   (1) ⌞7;U. L.⌟ A broad faint elevation.
   (2) ⌞5̇;<0.5⌟ Rather high elevation.
   (3) ⌞5̇;U. L.⌟ Faint mound.
   (4) ⌞4;U. L.⌟ A mound on the west wall.

Fontana

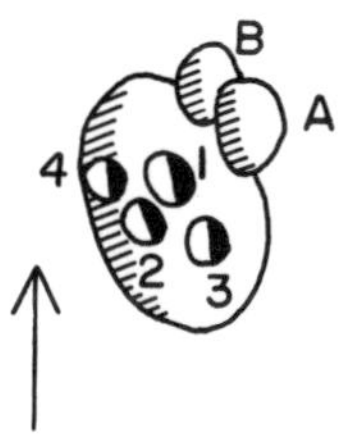

6. Fourier, U. F. to the S. of (-698,-534;78)
   This formation is generally regarded as a part of larger one, registered as 37504 (ξ =-0.705, η =-0.540) in the Arizona Catalogue.
   (1) ⌞6;0.4⌟ A low hill.
   (2) ⌞5.5;0.4⌟ Another low hill, probably the true central peak.
   (3) ⌞1.5;U. L.⌟ Very minute mound on the north bottom of the (2).
   (4) ⌞4;0.3⌟ A peak on the north part of the floor.
   (5) ⌞2.5;0.2⌟ A faint mound on the south bottom of the (1).

U. F. to the South of Fourier

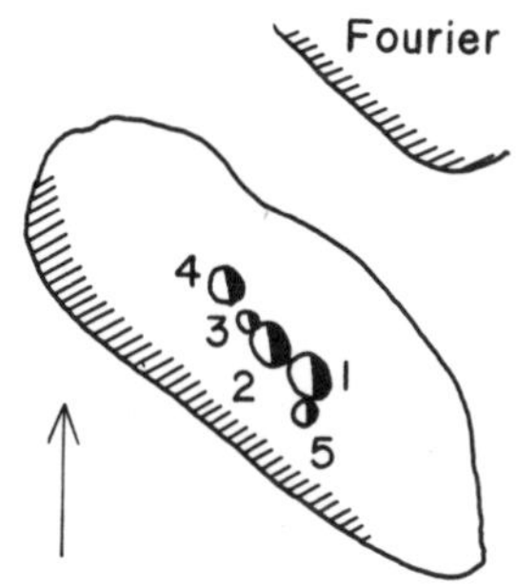

Gassendi, A

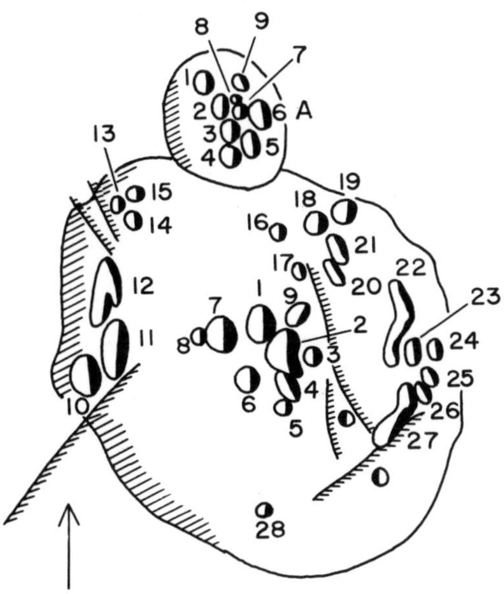

7. Gassendi (-611,-268;110)
   (1): β ⌞8;0.9⌟ A large peak.
   (2): ε ⌞11;0.8⌟ Lower, but larger mountain than the (1).
   (3) ⌞4;0.4⌟ Mound standing on the east of the (2).
   (4) ⌞4/3;0.4⌟ A peaked hill.
   (5) ⌞4;U. L.⌟ Faint mound.
   (6) ⌞5̇;0.4⌟ A small peak, but it may be the true central peak.

(7) ⌈10;0.7⌋ Large peak.

(8) ⌈4;U. L.⌋ Very faint mound.

(9) ⌈5̇/4;0.3⌋ A low hill.

(10) ⌈11;0.9⌋ A ridge, which is collapsed part of the west wall.

(11) ⌈14/5̇;0.3⌋ A mountain block, having a irregular form. The (8), (10) and (11) form a penetrating structure of the rim of Mare Humorum.

(12) ⌈14/6;0.4⌋ A hill having a form of fork.

(13) ⌈4;0.5̇⌋ (14) ⌈5/4;<0.6⌋ (15) ⌈6/4;<0.6⌋ Oblong mounds.

(16) ⌈4;0.4⌋ A small peak.

(17) ⌈4;0.4⌋ Another peak near a rill.

(18) ⌈4.5;<0.5⌋ A low hill.

(19) ⌈6;<0.4⌋ Faint plateau. (20) ⌈9.5/4;0.4⌋ (21) ⌈7/5̇;<0.8⌋ Oblong hills.

(22) ⌈20/4;<0.8⌋ A very long, curved hill.

(23) ⌈6/4;0.8⌋ (24) ⌈5̇/4;0.4⌋ (25) ⌈4;0.4⌋ (26) ⌈4;0.5̇⌋ Oblong hills.

(27) ⌈13/4;0.6e⌋ Bank along the rill VII.

(28) ⌈2.5;0.1⌋ An isolated minute mound.

8. Gassendi A (-616,-268;33)

(1) ⌈5.0;1.0⌋ Round mound on the north wall.

(2) ⌈6/5̇;0.8⌋ The most distinct peak.

(3) ⌈6/4;0.4⌋ An oblong hill.

(4) ⌈5̇;0.3⌋ Another mound on the south wall.

(5) ⌈8/4;U. L.⌋ Faint, oblong elevation.

(6) ⌈9;U. L.⌋ Very broad elevation.

(7) ⌈2.5;U. L.⌋ Small, faint elevation.

(8) ⌈2;U. L.⌋ Very tiny mound on the east bottom of the (2).

(9) ⌈5;U. L.⌋ Faint hill which may be collapsed part of the north-east wall.

9. Henry (Prosper) (-785,-400;42)

(1) ⌈2.5;U. L.⌋ A very faint mound.

(2) ⌈4;<0.2e⌋ A part of east wall of a small floor crater, whose diameter is about 5 km.

(3) ⌈3.5;0.6⌋ Low mound.

(4) ⌈8/5̇;<0.6⌋ Broad, oblong hill.

Henry (Prosper)

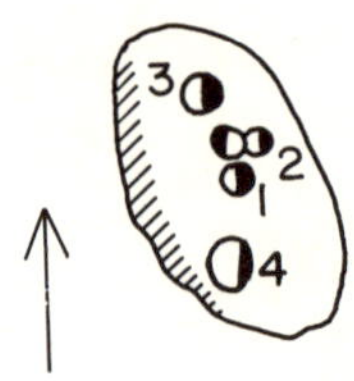

10. Lagrange (-795,-545;160)

(1) ⌈8;1.0w⌋ (2) ⌈10;0.6w⌋ (3) ⌈8;1.0⌋ (4) ⌈9;0.5̇⌋

(5) ⌈23/10;1.0⌋ These hills form a mountain block of irregular form.

(6) ⌈8;1.5̇⌋ (7) ⌈7;0.9⌋ Mountains on the south part of the floor.

(8) ⌈38/18;2.5⌋ Lofty hill which may be east wall of the craters Lagrange M and N.

(9) ⌈5̇;0.4⌋ A low hill.

(10) ⌈25/3;<1⌋ Long and uncinate dyke.

(11) ⌈6;U. L.⌋ Low hill.

(12) ⌈10;U. L.⌋ Faint elevation at the bottom of the east wall.

(13) ⌈13;U. L.⌋ Very broad and low elevation.

(14) ⌈8;0.5⌋ (15) ⌈8;0.5⌋ Very low uncinate hills, which are likely to be east

Lagrange

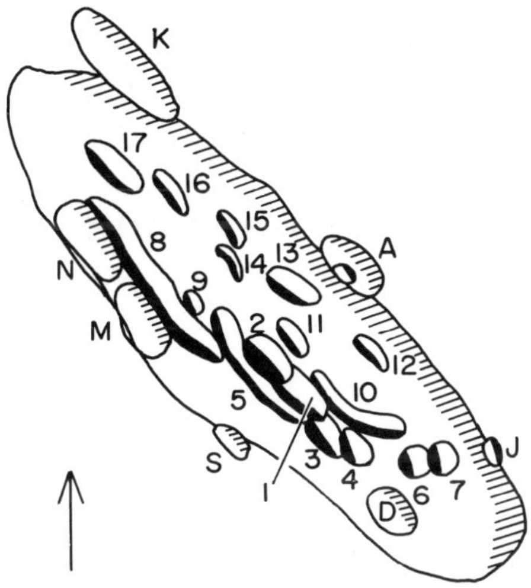

(XX)

walls of the shallow craters.

(16) ⌞11;0.7⌟ A low dyke which may be east wall of a shallow crater.

(17) ⌞12;0.6⌟ Large but low white dyke.

U.F. on the East of Lagrange E

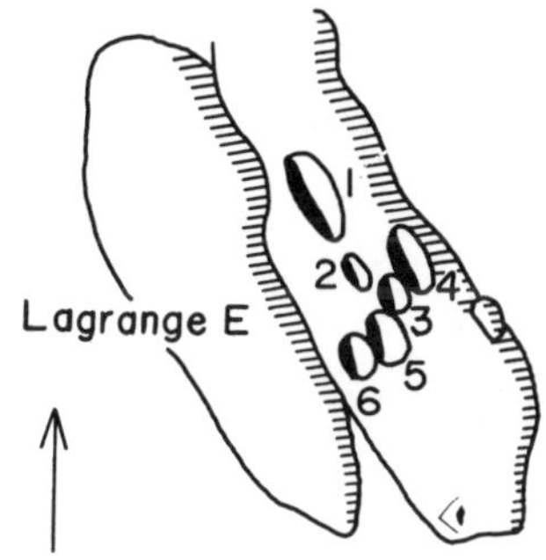

11. Lagrange E, U. F. on the E. of (-822,-486;57)
   (1) ⌞15;2⌟ A large mountain, of which north part is highest.
   (2) ⌞3;1⌟ A small hill attached to the south bottom of the (1).
   (3) ⌞3;1⌟ Another small hill.
   (4) ⌞11;1.5⌟ Mountain on the east part of the floor.
   (5) ⌞7;U. L.⌟ (6) ⌞5;U. L.⌟ Very faint hills.
   (7) ⌞5;U. L.⌟ Low mound.

12. Lagrange R (-829,-522;139)
   (1) ⌞13;2⌟ The central mountain.
   (2) ⌞10;1⌟ A hill on the west part of the floor.
   (3) ⌞17;2⌟ Large hill, which may be collapsed part of the north wall.

Lagrange R

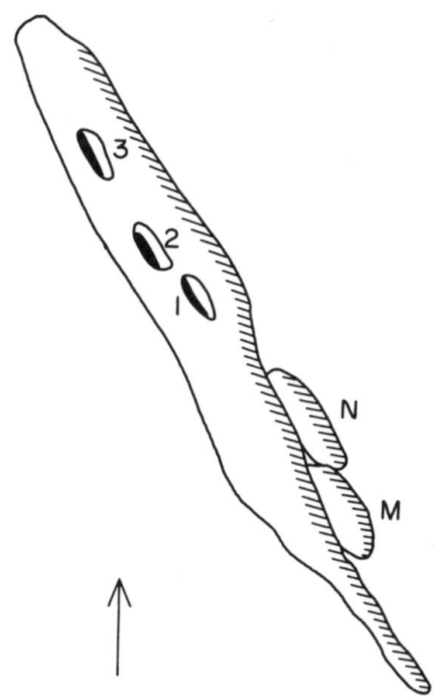

13. Lamarck A (-857,-423;56)

Lamarck A, D

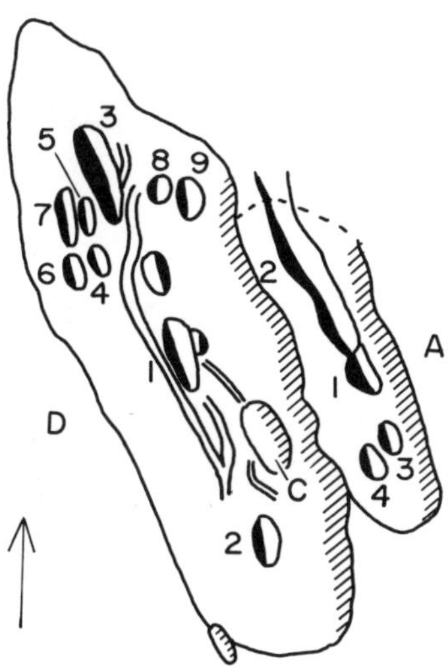

(1) ⌞8;2⌟ (2) ⌞35/20;4⌟ The central hill (1) is cinnected with a large mountain block, (2), which lies on the north part of the present formation.

(3) ⌞5;U. L.⌟ (4) ⌞10;U. L.⌟ Faint elevations on the south part of the floor.

(XX)

14. Lamarck D (-871,-425;109)
   (1) ⌊15;1⌋ The central peak.
   (2) ⌊7;U. L.⌋ Broad low mound on the south part of the floor.
   (3) ⌊20/10;2⌋ Long dyke which has only westerly slope.
   (4) ⌊5;2w⌋ (5) ⌊9;0.5w⌋ (6) ⌊4;U. L.⌋ (7) ⌊12;4w⌋ Hills standing on the north-west part of the floor.
   (8) ⌊4;U. L.⌋ (9) ⌊6;U. L.⌋ Very faint hills.

15. Mersenius D (-670,-392;34)
   (1) ⌊8;<0.8⌋ Broad elevation on the south part of the floor.
   (2) ⌊4;<0.8⌋ A hill standing on the west bank of the Rima Mersenius II, which runs through the present formation.

Mersenius D

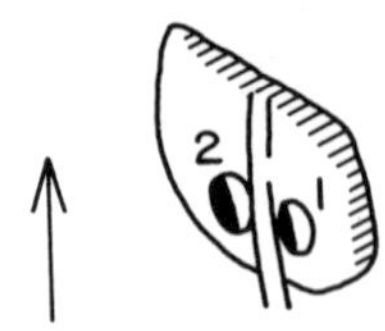

16. Mersenius P (-696,-341;42)
   (1) ⌊4;0.4⌋ Round mound.
   (2) ⌊2.5;0.3⌋ Small peak.
   (3) ⌊1.5;<0.3⌋ Minute mound on the bottom of the south-west wall.
   (4) ⌊7;0.6⌋ A low heap on the south wall.

Mersenius P

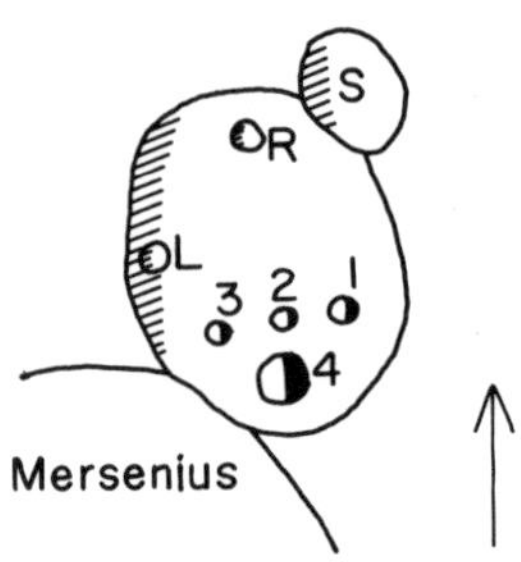

17. Palmieri A (-633,-534;19)
   (1) ⌊3;U. L.⌋ Very faint mound on the center of the crater.

18. Piazzi (-748,-590;101)
   (1) ⌊14/11;0.8⌋ A low, broad hill. It may be collapsed remnant of central peak.
   (2) ⌊14/5;0.3⌋ Smaller hill on the west side of the (1).
   (3) ⌊11/5;U. L.⌋ A part of terrace structure on the east wall.
   (4) ⌊3.5;0.3⌋ (5) ⌊5;U. L.⌋ Very faint elevations on southern part of the floor.
   (6) ⌊6;U. L.⌋ A faint elevation.
   (7) ⌊30/5;0.2⌋ (8) ⌊6;0.8⌋ The (7) is a dyke, on which stands a hill (8). The dyke (7) is a part of radial structure from the Mare Orientale, according to the Lunar Orbiter IV Photographs.

Piazzi

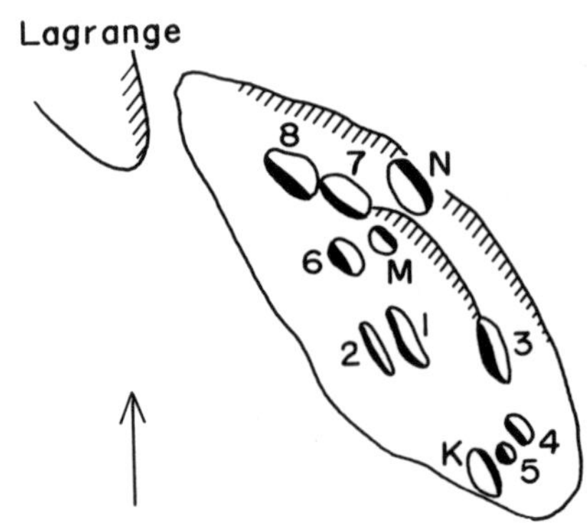

19. Rocca A (-914,-240;60)
   (1) ⌊11;0.3w⌋ The central mountain superposes on the north wall of the Rocca M.

20. Rocca E (-916,-206;46)
   (1) ⌊10;<0.3⌋ Faint elevation on the westerly eccentric position of the floor.
   (2) ⌊2.5;<0.2⌋ A small peak.

Rocca E

(3) [25/5;<0.5] Long faint ridge, but its northern end is very distinct.

21. Rocca M (-914,-252;35)

(1) [7;0.3] A low hill.

(2) [4;0.4] A distinct hill.

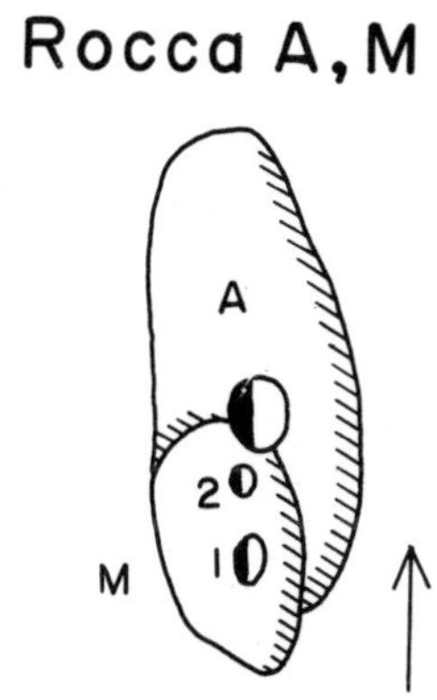

22. Rocca Z (-929,-281;39)

(1) [7;U. L.] (2) [6;U. L.] On both sides of a faint crack, stand two low hills.

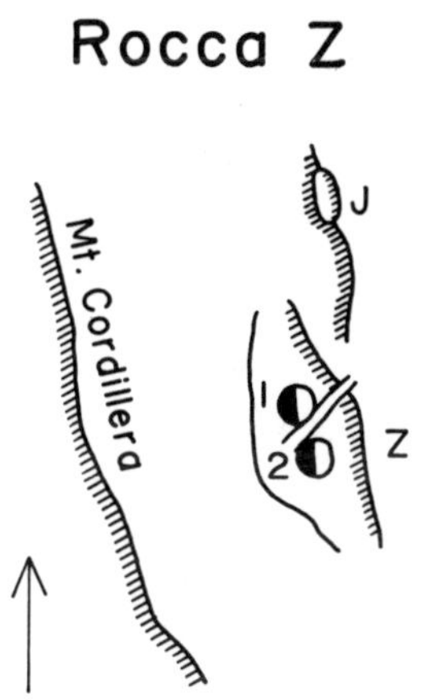

23. Sirsalis (-848,-216;44)

(1) [5.5;1.0̇] A round mound at the center of the floor. According to the Lunar Orbiter IV Photographs, it has a hill of complicated form, which looks like collapsed large mountain.

24. Vieta (-726,-488;50)

The rugged northern part of the floor forms a platform.

(1) [4;0.9] Central mound, connected with the west wall of Vieta H.

(2) [7;0.4] A gentle hill on the bottom of the south wall.

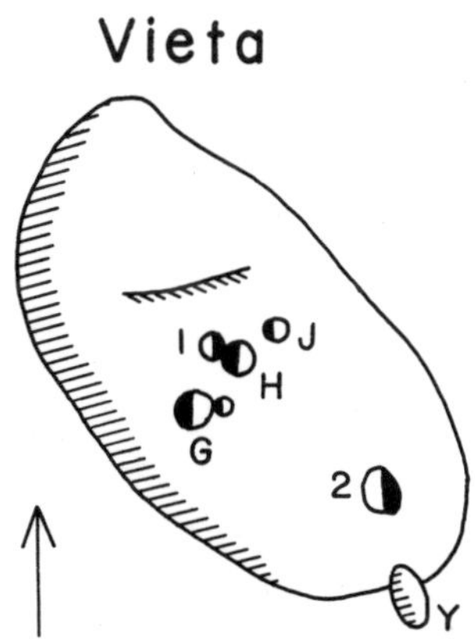

XXI

1. Bouvard (-761,-640;179)
    (1) ⌞29;0.3⌟ A large, but low hill.
    (2) ⌞20;2⌟ Peaked mountain on the northerly eccentric position of the floor.
2. Bouvard M (-641,-651;69)
    (1) ⌞7;<0.5⌟ A round mountain.
    (2) ⌞2.5;U. L.⌟ (3) ⌞2.5; U. L.⌟ Small mounds on the north of the (1).
3. Inghirami (-630,-737;91)
    The floor is very undulatory.
    (1) ⌞6;1.2⌟ (2) ⌞3.5;0.4⌟ These hills have forms of short dyke. The (1) is higher, and may be the true central peak.
    (3) ⌞4;0.3⌟ (4) ⌞5̇;<1.0⌟ Faint dykes, which may be the east bank of shallow depression.
    (5) ⌞15̇;1.2e⌟ A low elevation.
    (6) ⌞18;1.0e/1.7w⌟ A ridge.
    (7) ⌞6;0.5̇⌟ (8) ⌞4;0.3⌟ Faint ridges. Between the (6) and (7), there is a deep depression.
4. Inghirami Q (-641,-742;39)
    (1) ⌞8;U. L.⌟ A broad but very low central mountain.
5. Lacroix L, U. F. on the N. of (-699,-665;35/20)
    (1) ⌞3;U. L.⌟ A faint central mound.

Bouvard, M

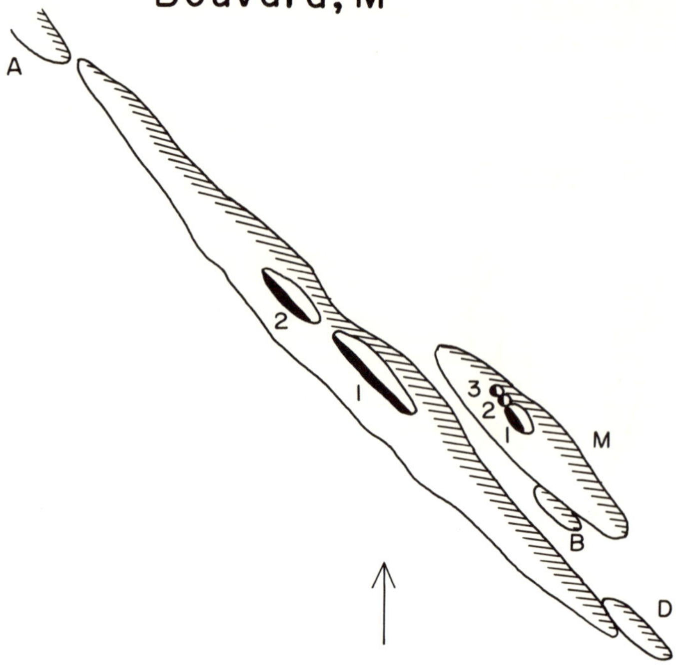

Inghirami, Q

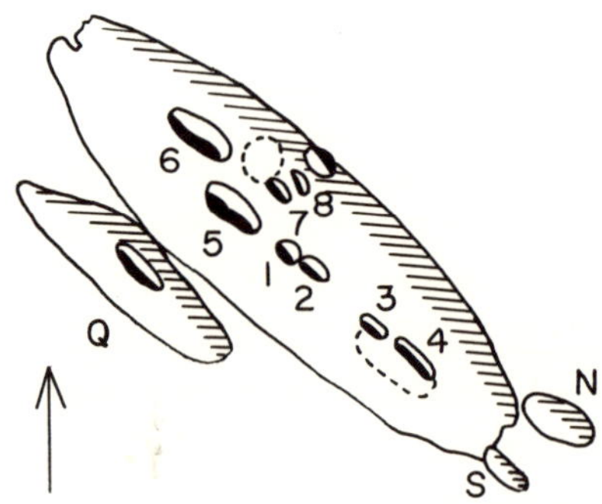

XXII

1. <u>Bailly, U. F. to the N.-W. of</u> (-491,-869;62)
   This formation was formerly called as Pingré S.
   (1) ⌜4;U. L.⌝ A round central mound on rugged floor.
2. <u>Bettinus</u> (-316,-895;71)
   (1) ⌜10̇;1.2⌝ A central peak having a conical form.
3. <u>Brown E</u> (-207,-729;22)
   (1) ⌜2.0;0.5̇⌝ A small central peak.
   (2) ⌜4;0.3⌝ Broad and low mound on the bottom of the south wall.
   (3) ⌜4;0.3⌝ Low mound on the south wall.

Brown E

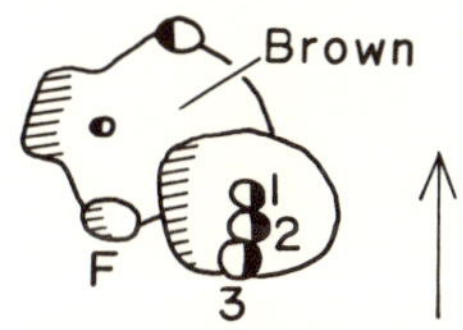

4. <u>Epimenides</u> (-380,-655;27)
   (1) ⌜8;<0.6⌝ A centrally elevated floor. On the wall, a faint terrace is seen. It seems as a double crater.
5. <u>Epimenides A, U. F. to the N. -E. of</u> (-355,-679;15)
   (1) ⌜5/2;0.1⌝ A faint linearment. It may be the east wall of a small crater, superposing on the west wall of the present formation.
6. <u>Epimenides S</u> (-366,-664;26)
   (1) ⌜7;1.1⌝ Large mound.
   (2) ⌜4;<0.3⌝ A low mound on the south wall.
   (3) ⌜1.5;0.2⌝ Very minute mound on the south of the (2).
   (4) ⌜3;0.5⌝ Faint mound on the south-west wall.
   (5) ⌜4;U. L.⌝ A broad elevated part on the bottom of the west wall.

Epimenides S

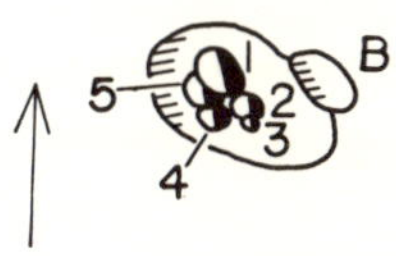

7. <u>Haidinger</u> (-328,-631;21)
   (1) ⌜3;0.1⌝ Very faint central mound.
   (2) ⌜2.5;U. L.⌝ A low mound, which may be the east wall of a very minute crater, whose diameter is about 3km.
   (3) ⌜4;U. L.⌝ Broad elevation on the bottom of the north wall.

Haidinger

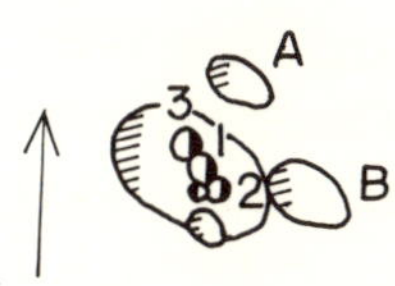

8. <u>Haidinger C</u> (-293,-630;19)
   (1) ⌜4;U. L.⌝ A low mound on the north-westerly eccentric position of the floor.
9. <u>Haidinger M</u> (-297,-608;22)
   (1) ⌜2;0.4⌝ A very minute central peak.
10. <u>Hainzel</u> (-415,-660;70)
   (1) ⌜6;0.4⌝ A large peak.
   (2) ⌜6;0.3⌝ Elevated part of the floor.
   (3) ⌜7;0.5̇⌝ Mountain block, on whose north side, a crease is seen.
11. <u>Hainzel A</u> (-426,-646;53)
   (1) ⌜6;1.0⌝ Eminent central peak.
   (2) ⌜6;0.9⌝ Round mound on the bottom of the north wall.
   (3) ⌜7/4;1.0e⌝ Oblong hill.
   (4) ⌜7;0.8⌝ Low mountain.
12. <u>Hainzel B</u> (-434,-615;15)

(XXII)

(1) ⌊6;U. L.⌋ This formation is a double crater. A small, concentric faint circle is seen.

13. Hainzel C (-408,-657;38)
 (1) ⌊5;0.7⌋ Eminent central peak.
 (2) ⌊3;0.1⌋ Small attached hill of the (1).
 (3) ⌊3;0.2⌋ Round mound on the north wall.
 (4) ⌊4;0.2⌋ Low mountain.

14. Hainzel R (-463,-626;19)
 (1) ⌊3;U. L.⌋ A faint hill. This formation may be a compound crater, and the hill may be south-east wall of the superposed crater.

15. Heinsius (-234,-636,66)
 (1) ⌊4;U. L.⌋ A faint mound. The west part of this formation is superposed by a crater, whose diameter is about 40 km. The mound is central peak of this crater.

16. Heinsius A (-232,-636;20)
 (1) ⌊7/4;<0.2⌋ A low and broad elevation lies from the south wall to the center.

17. Heinsius B (-245,-642;24)
 (1) ⌊4;0.6⌋ A hill on the north wall.

18. Heinsius C (-233,-651;22)
 (1) ⌊4;0.5⌋ A round mound on the south wall.

19. Heinsius MA (-203,-647;23)
 (1) ⌊4.5;U. L.⌋ A very low, but broad mound.

20. Kircher A (-272,-914;29)
 (1) ⌊10;1.1⌋ A large central mountain.

21. Kircher D (-292,-924;39)
 (1) ⌊5;0.8⌋ A round mound is seen on the boundary of the neighboring unnamed crater, whose diameter is 26 km.

22. Lagalla (-273,-702;85)
 (1) ⌊3;0.5⌋ An uncinate peak.
 (2) ⌊7/3;0.5⌋ A long dyke, whose northern end is highest.
 (3) ⌊7/3;0.4⌋ An oblong hill.
 (4) ⌊5/4;0.3⌋ Two-peaked hill.
 (5) ⌊8;0.2⌋ Triangular plateau.
 (6) ⌊4;0.3⌋ A mound.
 (7) ⌊4;0.5⌋ Round mound on the bottom of the south wall.

23. Longomontanus (-240,-761;145)

Hainzel, A,C

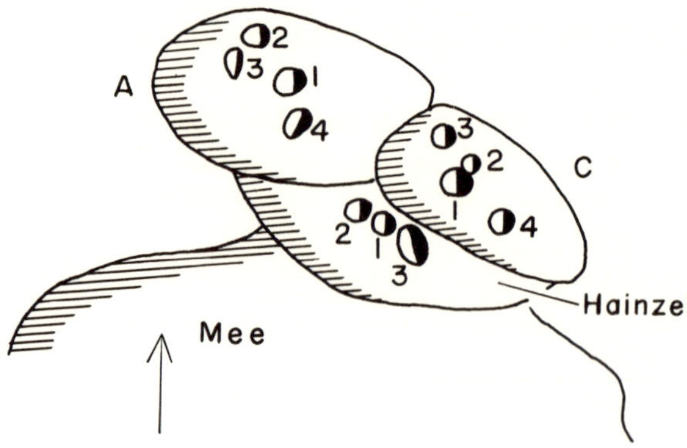

Heinsius, A,B,C

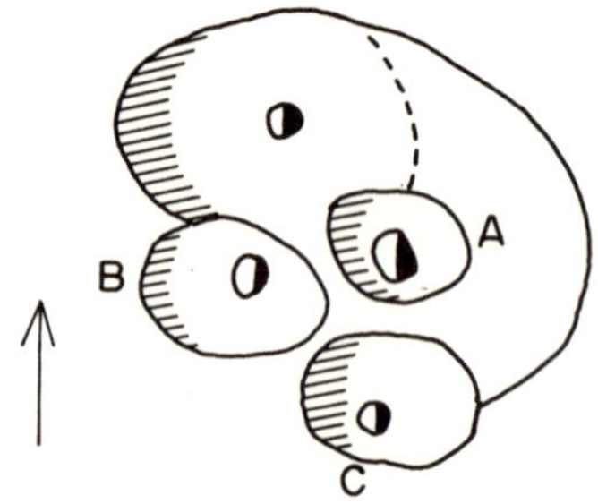

Lagalla

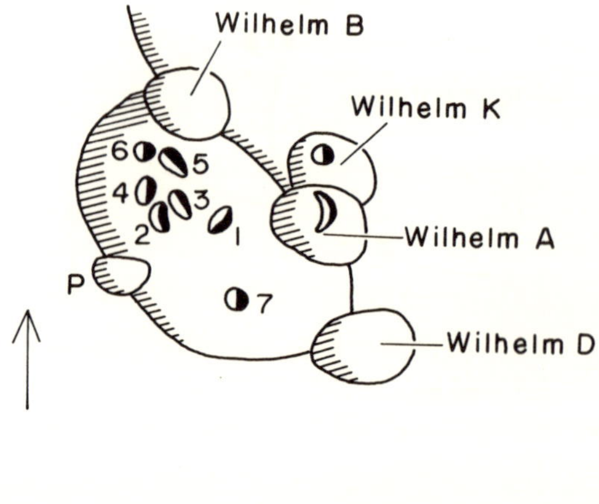

All ridges and linearments are parallel to the structure around this formation.

(1) [5.5;0.9] A large mountain block.

(2) [5.0;0.8] A hill lower than the (1).

(3) [5.0;0.8] A short ridge.

(4) [2.5;0.1] (5) [2.5;0.2] Faint hills.

(6) [3;0.1] (7) [3;U. L.] Faint linearments.

(8) [3;0.1] An isolated linearment.

(9) [9;U. L.] Low mountain block, on which we can perceive a faint linearment.

(10) [3;0.2] A mound.

(11) [2.5;0.2] A distinct mound.

(12) [12/2;0.2] (13) [6/3;0.6w] Long uncinate hills, which are parts of the wall of a shallow crater.

24. Longomontanus A (-246,-798;29)

(1) [4;0.8] A two-peaked mountain.

25. Longomontanus B (-213,-797;48)

(1) [7/5;0.4] An oblong, two-peaked hill.

(2) [3;0.4] Round mound.

26. Longomontanus G (-209,-751;15)

Longomontanus, G

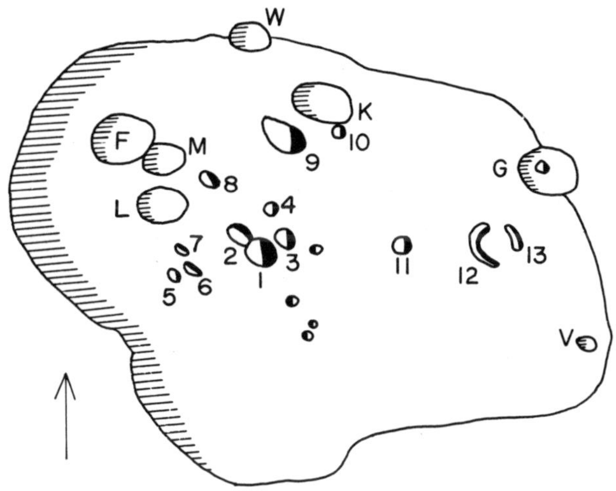

Longomontanus B

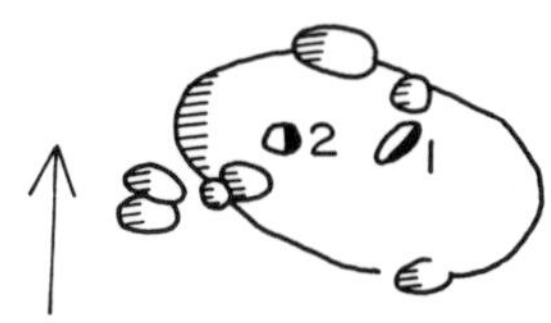

Longomontanus Z

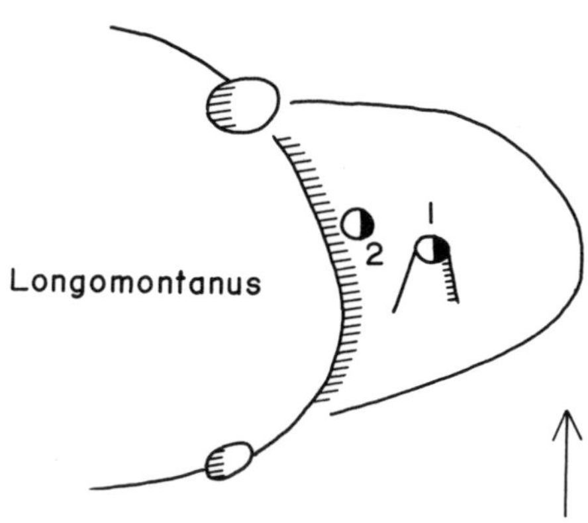

Mee

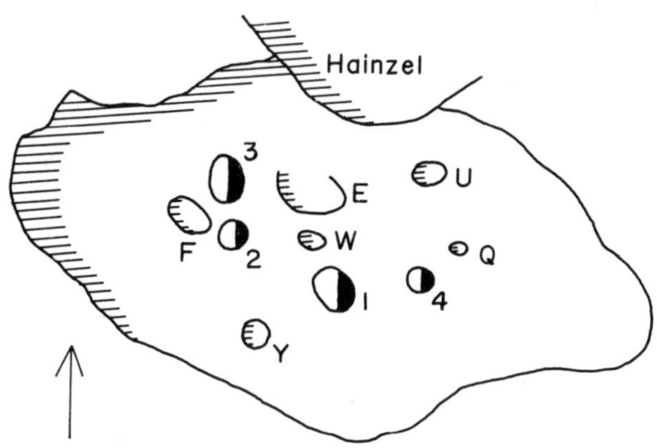

(1) [2;0.4] Very minute central peak.

27. Longomontanus Z (-206,-766;95)

(XXII)

(1) ⌞6;0.5⌟ A round hill, which stands on the north end of a low plateau.

(2) ⌞7;0.4⌟ A low peak.

28. Mee (-415,-692;132)

(1) ⌞13;1.4⌟ A large central mountain.

(2) ⌞7;1.4⌟ A mountain forming the east wall of Mee F.

(3) ⌞6;1.0̇⌟ A hill having easterly steep slope.

(4) ⌞6;0.8⌟ A distinct peak.

Scheiner

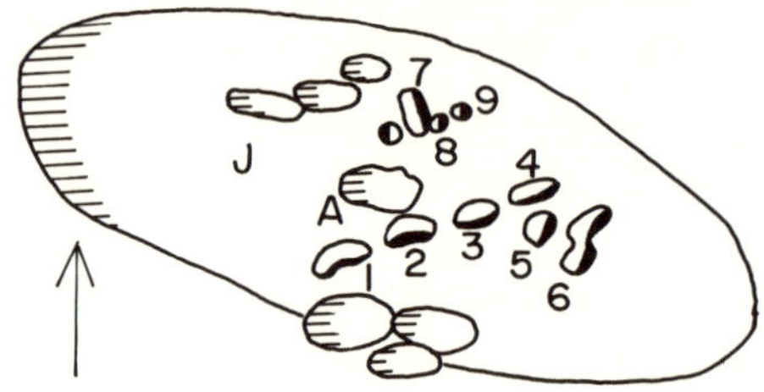

29. Scheiner (-230,-870;110)

(1) ⌞7/3;0.5⌟ A ridge, which is the extension of the collapsed part of the south wall.

(2) ⌞10/3;0.2⌟ Two-peaked hill.

(3) ⌞6/4;U. L.⌟ (4) ⌞4; U. L.⌟ (5) ⌞5;U. L.⌟ Low hills.

(6) ⌞10/3;U. L.⌟ A curved hill.

(7) ⌞8/4;0.3⌟ An oblong hill.

(8) ⌞1.5;U. L.⌟ (9) ⌞1.5 ;U. L.⌟ Tiny mounds. The (1), (2), (3) and (4) stand on a line, and the (8) and (9) stand on another line, which are parallel to each other.

Schickard

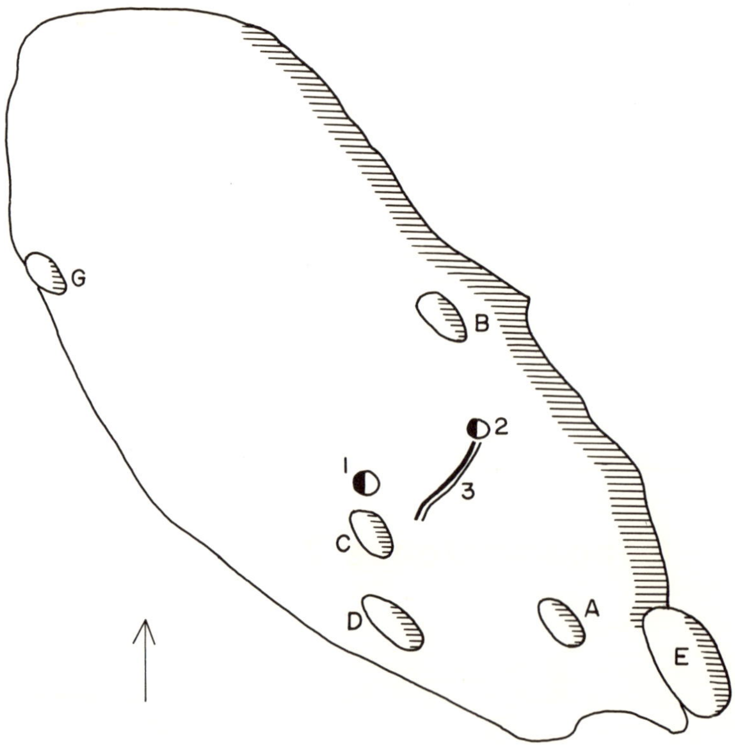

30. Scheiner B (-278,-862; 29)

(1) ⌞5̇;1.1⌟ A round central mound. The slope of the wall is steeper on the side towards the center.

31. Schickard (-582,-700;227)

(1) ⌞5̇;1.1⌟ An isolated peak.

(2) ⌞4;0.2w/0.4e⌟ A peak lower than the (1).

(3) ⌞35/3;U. L.⌟ A long dyke between the (2) and Schickard C, whose east end is highest. The floor is somewhat lower on southern side of this dyke than on the northern side.

32. Schiller (-397,-786;178)

(1) ⌞25/10;2.0⌟ A mountain block having an irregular form.

(XXII)

Schiller

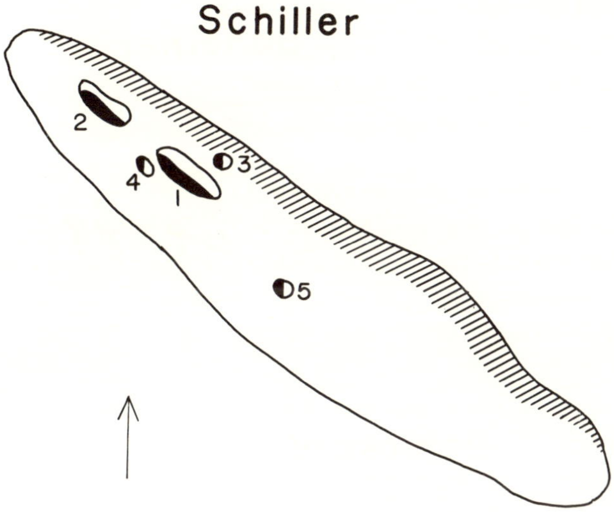

Wargentin

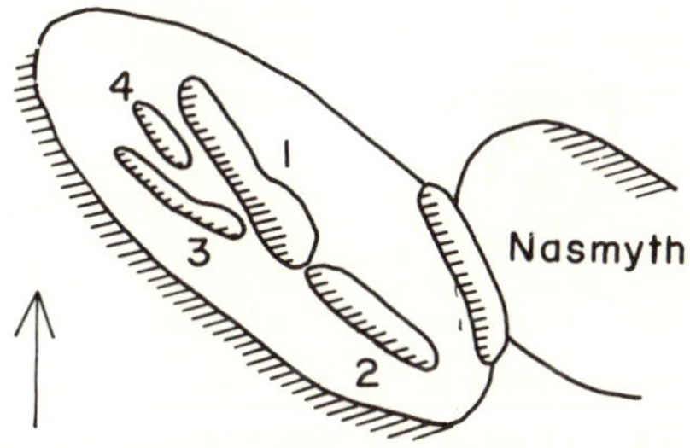

U.F. to the South of Schiller

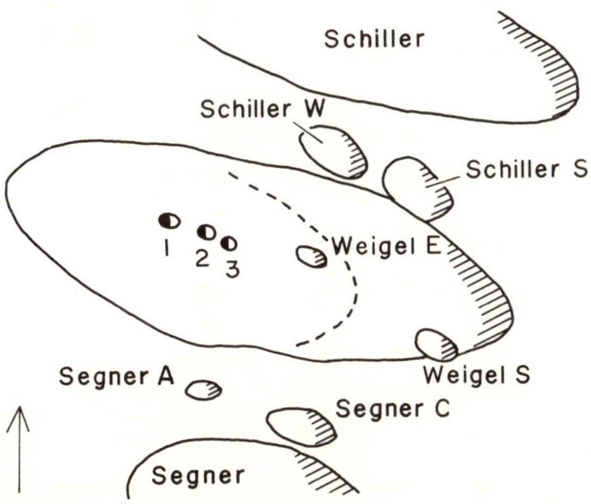

(2) ⌊12;2.0⌋ Another mountain block.

(3) ⌊2.5;0.2⌋ Small mound between the (1) and the east wall.

(4) ⌊5;1.5⌋ A large mound.

(5) ⌊3;0.2⌋ An isolated peak at the bottom of the west wall.

33. <u>Schiller, U. F. to the S. of</u> (-390,-823;120)

This formation is not registered in the Arizona Catalogue.

(1) ⌊4;0.3⌋ (2) ⌊3;0.2⌋ (3) ⌊2;0.1⌋ Low hills standing in a line.

34. <u>Segner</u> (-386,-856;67)

(1) ⌊5;U. L.⌋ The central part of the rugged floor is very faintly elevated.

35. <u>Wargentin</u> (-565,-760;84)

(1) ⌊40/7;U. L.⌋ (2) ⌊20/7;U. L.⌋ (3) ⌊20/10;U. L.⌋ (4) ⌊15/10;U. L.⌋ Very faint and broad dykes run radially on the plateau-like floor. The (1) is largest, and its south end, the center of the floor, is faintly elevated. The (2) runs on the southern part of the floor. The (3) is west radius. The (4) is the smallest.

36. <u>Wargentin F</u> (-569,-783;20)

(1) ⌊4;U. L.⌋ This formation is a compound crater, and the ridge in the crater is west wall of the larger component.

37. <u>Wilhelm A</u> (-267,-703;20)

Configuration of this formation is shown in the figure of Lagalla.

(1) ⌊6/2;U. L.⌋ This formation is a compound crater, and the faint dyke (1) is

the east wall of the smaller component.

Wilhelm D

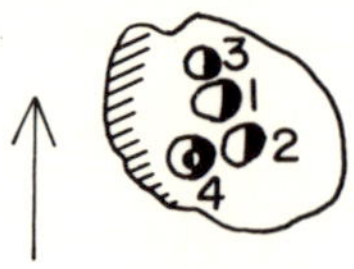

38. Wilhelm D (-227,-667;32)
   (1) ⌞4.5;U. L.⌟ Very faint central mound.
   (2) ⌞4.5;0.5̇⌟ A distinct mound at the bottom of the south wall.
   (3) ⌞5/2;U. L.⌟ A dyke on the north wall.
   (4) ⌞5̇;0.6⌟ A mound on the south-west wall, on which we can perceive a hill-top crater, whose diameter is 2 km.

39. Wilhelm GA, U. F. on the N. of (-315,-664;27/24)
   (1) ⌞4;0.4⌟ A low ridge on the rugged floor.

40. Wilhelm K (-265,-696;21)
   (1) ⌞6;0.4⌟ The west half of the floor is elevated and forms a plateau. The east edge of the plateau is a part of a structure through this formation.

41. Wilhelm V (-241,-693;8)
   (1) ⌞1.3;U. L.⌟ A very minute central mound.

42. Zuchius (-370,-879;64)
   (1) ⌞6;0.6⌟ A distinct peak.
   (2) ⌞4;0.3⌟ Small mound, which may be the true central peak.
   (3) ⌞6;0.8⌟ A ridge on the eastern part of the floor.

Zuchius

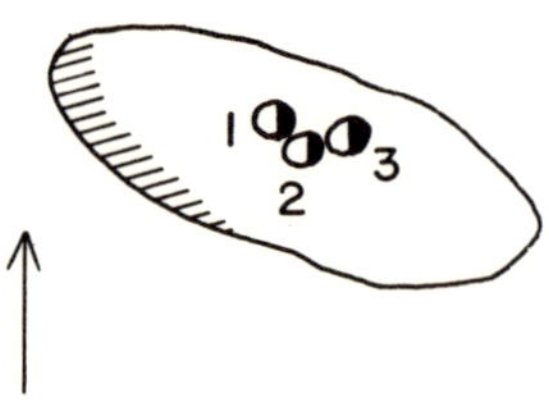

XXIII

1. Amundsen (+070,-997;65)
   (1) ⌜15;U. L.⌝ A central mountain having three or more peaks, according to photographs of the Lunar Orbiter IV.
2. Baco K (+178,-809;29)
   (1) ⌜6;U. L.⌝ Faint mound on the northerly eccentric position of the floor.
3. Ball C (-120,-611;31)
   (1) ⌜8;0.7⌝ A large central mountain.

Blancanus

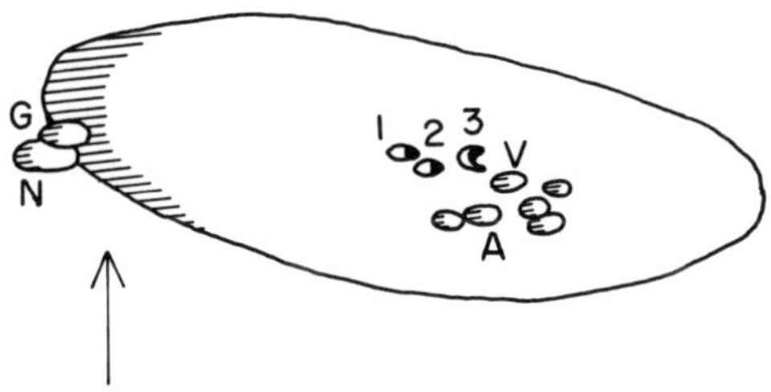

4. Blancanus (-163,-896;105)
   (1) ⌜4;0.3⌝ A peak.
   (2) ⌜5̇;0.4⌝ Another lager peak.
   (3) ⌜5;0.3⌝ A part of a small crater.
5. Blancanus C (-187,-917;46)
   (1) ⌜5;0.5⌝ A ridge at the bottom of the west wall.
6. Boguslawsky N, U. F. on the N. of (+161,-959;42)
   (1) ⌜5;U. L.⌝ A low mound on the bottom of the north wall of Boguslawsky N.
7. Casatus A (-184,-956;50)
   (1) ⌜6;0.5⌝ A central mound.

U.F. on the North of Boguslawsky N

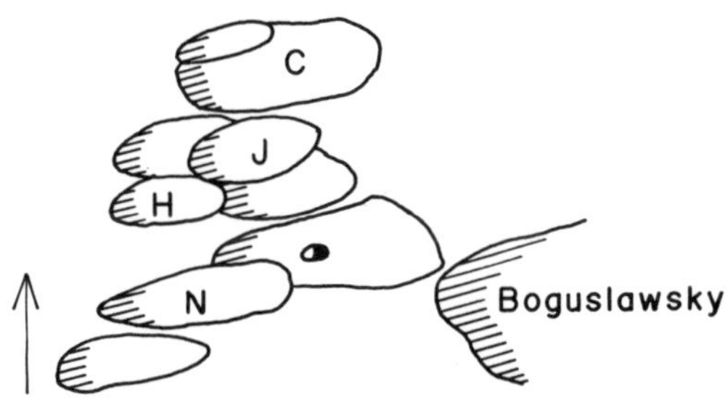

Clairaut B

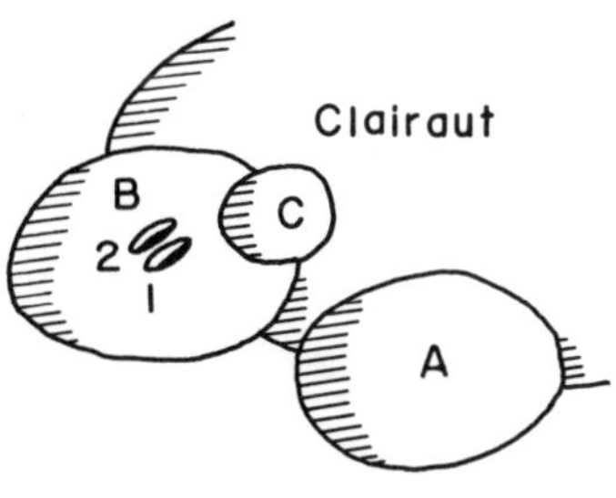

8. Clairaut B (+145,-747;43)
   (1) ⌜5/3;U. L.⌝ (2) ⌜5/3;U. L.⌝ Very faint parallel ridges.
9. Clavius (-130,-852;225)
   (1) ⌜8;0.6⌝ A mountain with a hill-top crater, which is very shallow.
   (2) ⌜6;0.2⌝ A low ridge.
   (3) ⌜5;0.6⌝ A mountain, which has a shallow crater near its summit.
   (4) ⌜4;0.8⌝ An isolated hill.
   (5) ⌜11/3;0.2⌝ Faint scarp which has slope towards east.
   (6) ⌜2;0.3⌝ (7) ⌜2;0.3⌝ Small peak on a ridge.
   (8) ⌜5̇/3;U. L.⌝ An extension of the above ridge.
   (9) ⌜10/2;U. L.⌝ (10) ⌜10/3;U. L.⌝ Faint ridges.
   (11) ⌜50/3;0.3⌝ (12) ⌜30/3;0.3⌝ (13) ⌜50/3;0.3⌝ (14) ⌜30/4;0.3⌝ Parallel dykes

Clavius, D, K, L, Porter, Rutherfurd

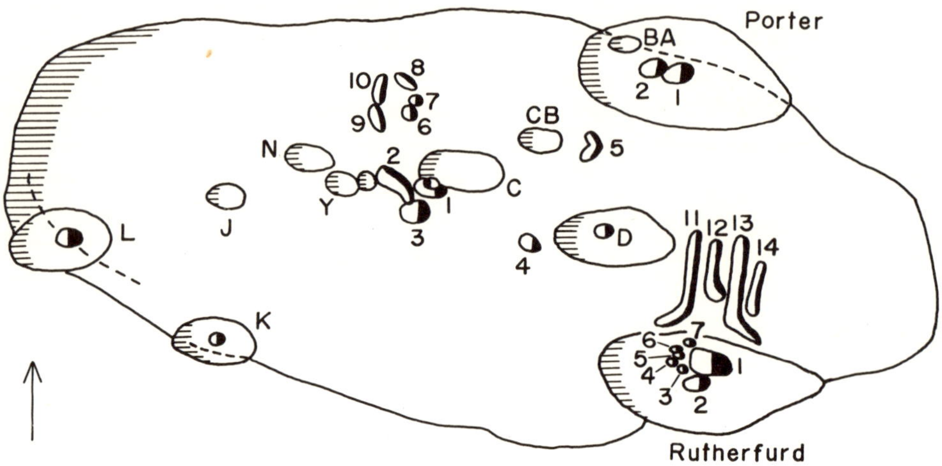

from Rutherfurd.

10. Clavius D (-110,-854;28)

    (1) ⌈5̇;0.5⌋ A low mound on the westerly eccentric position of the floor.

11. Clavius K (-166,-869;20)

    Extension of the wall of Clavius passes through this formation.

    (1) ⌈3;0.5⌋ A low central mound.

12. Clavius L (-188,-854;24)

    (1) ⌈5;0.5⌋ A broad mound on the south wall, which is a part of continuation of Clavius' wall.

Curtius

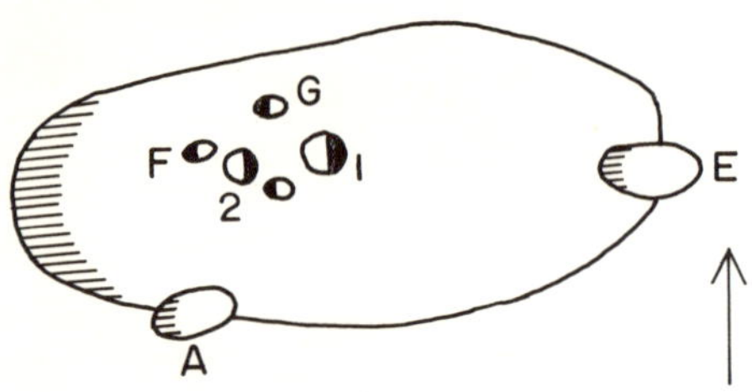

13. Curtius (+030,-922;95)

    (1) ⌈6;0.6⌋ A peaked mountain, whose west side is slightly depressed, on the westerly eccentric position of the floor.

    (2) ⌈5̇;0.2⌋ A low hill.

14. Curtius B (+036,-896; 41)

    (1) ⌈5̇;<0.3⌋ Faint mound.

    (2) ⌈4;0.1⌋ Faint mound on the bottom of the south wall.

Curtius B

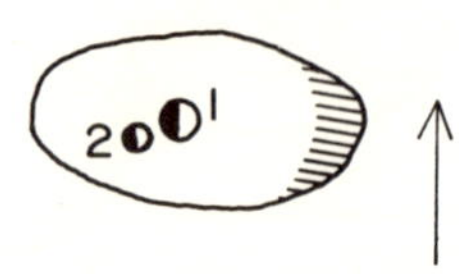

Curtius D

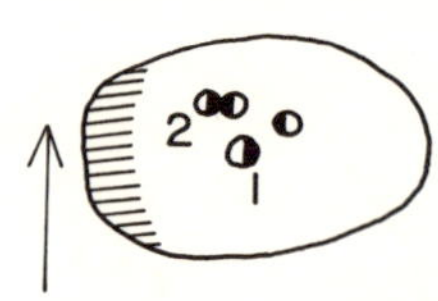

15. Curtius D (+060,-905;61/45)

    (1) ⌈4;U. L.⌋ Faint hill.

    (2) ⌈3.5;0.3e⌋ A peak, which is west wall of a small crater.

16. Cuvier A (+127,-792;18)

    (1) ⌈5̇;U. L.⌋ A faintly centrally elevated floor.

17. Cuvier B (+148,-784;17)
(1) [3;U. L.] Centrally elevated floor.

18. Cuvier E (+137,-791;19)
(1) [10/3;U. L.] An oblong elevation from the center to the edge of the south wall.

19. Cuvier O (+130,-784;10)
(1) [2.5;U. L.] A small cemtral mound.

20. Cuvier Q (+114,-783;13)
(1) [3;U. L.] A faint central mound.

Cysatus

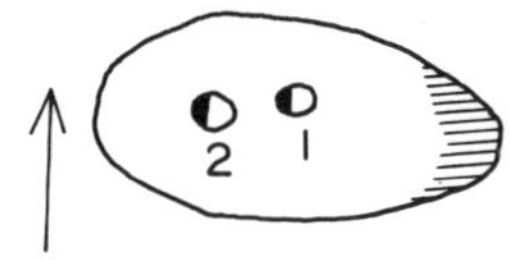

21. Cysatus (-043,-914;49)
(1) [4;U. L.] A very faint central mound.
(2) [4;U. L.] A faint mound on the south-west wall.

Deluc, D

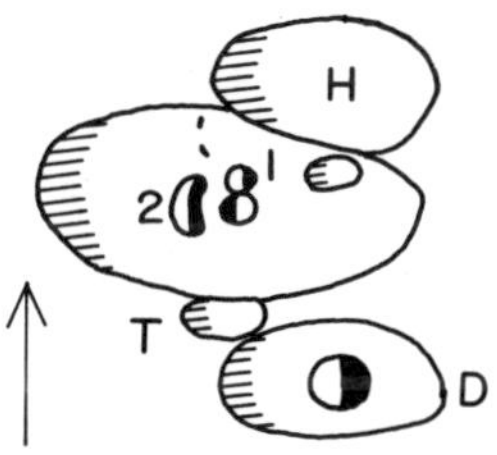

22. Deluc (-028,-819;47)
(1) [3;0.3] A small peak.
(2) [10/4;U. L.] Broad faint dyke.

23. Deluc A (-004,-810;55)
(1) [7;0.2] A broad and faint mound.

24. Deluc AB (+006,-823;22)
(1) [7/3;U. L.] (2) [5̇/3;U. L.] Parallel dykes on the western part of the floor, running from south to north.

Deluc AB

25. Deluc D (-023,-832;27)
(1) [7;U. L.] A faint elevation on the easterly eccentric position of the floor.

26. Deluc H (-021,-810;26)
(1) [12;U. L.] A faint and broad elevation.

27. Deluc J, U. F. on the E. of (+014,-895;7)
(1) [1.5;U. L.] A small mound is visible under illumination from east.

28. Demonax (+175,-979;114)
(1) [8;1] A large central peak.

29. Drygalski (-180,-982;176)
(1) [20;2] A very large central mountain.

Demonax

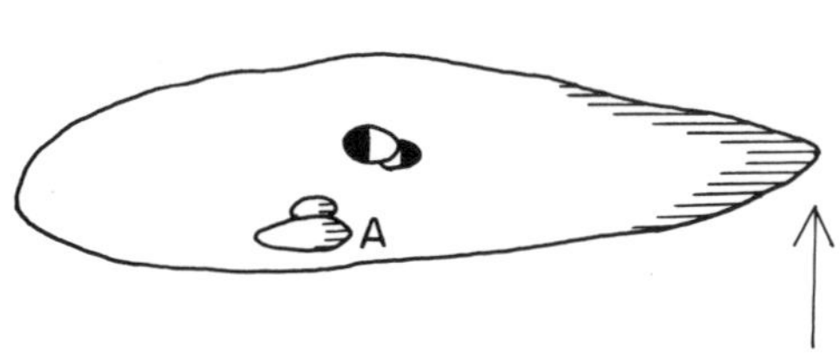

30. Faraday (+112,-675;69)
(1) [2;0.2] Elevated west wall of a small crater.
(2) [2;0.3] A small mound.

Faraday, C

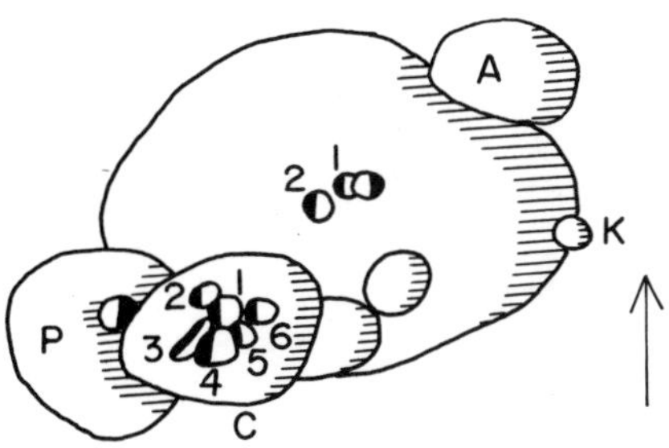

31. Faraday C (+102,-685;30)
(1) [5̇;0.5̇] A distinct peak on the northerly eccentric position of the floor.
(2) [2;0.3w] A peak at the bottom of the north

wall.

(3) ⌊7/1;5;0.1⌋ A faint dyke.

(4) ⌊6;0.1⌋ Broad elevation.

(5) ⌊4;U. L.⌋ A faint elevated part at the bottom of the south-east wall.

(6) ⌊3;0.1⌋ Rather distinct mound at the bottom of the east wall.

32. Faraday G (+122,-718;31)

(1) ⌊6;1.2w/0.4e⌋ A central mountain of irregular form, on somewhat southerly eccentric position.

33. Faraday L, U. F. to the N.-E. of (+122,-619;8)

(1) ⌊2;U. L.⌋ A very small mound on the southerly eccentric position of this crater. It may be part of the north wall of neighboring crater.

34. Gemma Frisius Y, U. F. on the S.-E. of (+193,-612;6)

(1) ⌊2.5;0.1⌋ A small round mound.

35. Gruemberger (-068,-920;94)

(1) ⌊4;0.3⌋ A small peak.

(2) ⌊14/4;0.2⌋ A long hill.

(3) ⌊2.5;0.2⌋ Very small mound.

(4) ⌊3.5;0.2⌋ A small mountain.

(5) ⌊3.0;0.2⌋ Very faint hill.

(6) ⌊10/2;0.1⌋ A low scarp having easterly steep slope.

Gruemberger

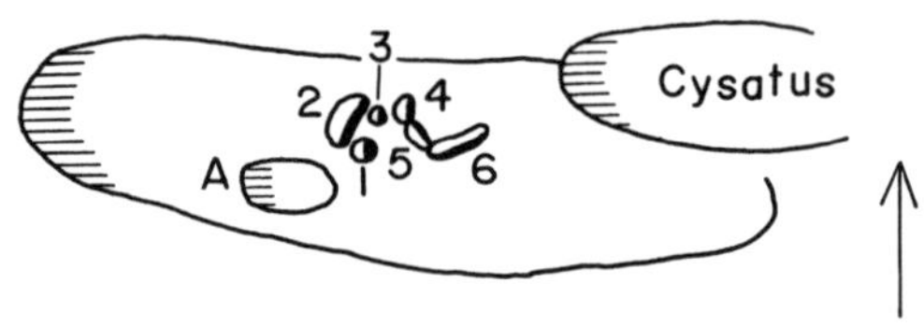

36. Heinsius M (-199,-656;13)

(1) ⌊2.0;0.3⌋ A small but distinct mound on the bottom of the south-west wall.

Heinsius M, MA, P, Q

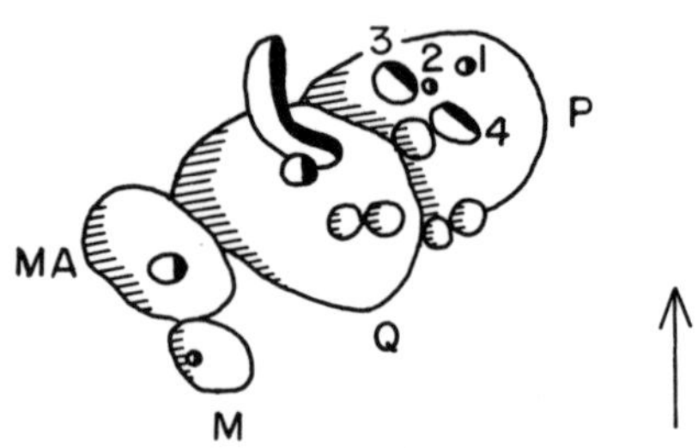

37. Heinsius P (-182,-635;37)

(1) ⌊2.5;0.4e⌋ A small peak on the north-east wall.

(2) ⌊2;0.6⌋ A small mound on the lowest part of the floor.

(3) ⌊6;U. L.⌋ (4) ⌊5;0.6⌋ Broad and low mountain blocks.

38. Heinsius Q (-191,-642;37)

(1) ⌊4;0.4⌋ A round low mound.

39. Heraclitus (+071,-757;90)

(1) ⌊30/10;0.8⌋ A large ridge.

(2)⌊3;0.2⌋ (3) ⌊3;0.4⌋ (4) ⌊3;0.3⌋ (5) ⌊2.5;0.2⌋ Small peaks on the west part of the floor.

(6) ⌊4;0.4⌋ An isolated peak on the east part of the floor.

The east half of this formation forms a crater, whose west wall is the ridge (1), and the (6) may be its central peak.

Heraclitus, D

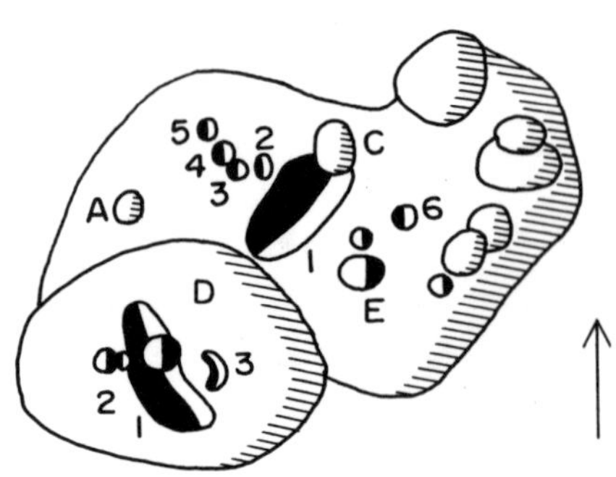

40. Heraclitus D (+058,-770;52)

(1) ⌊11;0.9⌋ A large mountain block, which has a

shallow crater on its north bottom.

(2) ⌞3;0.2w⌟ A small mound, which forms the east wall of a small crater.

(3) ⌞6/2;0.2⌟ An uncinate hill, and forms east wall of a shallow crater.

Huggins

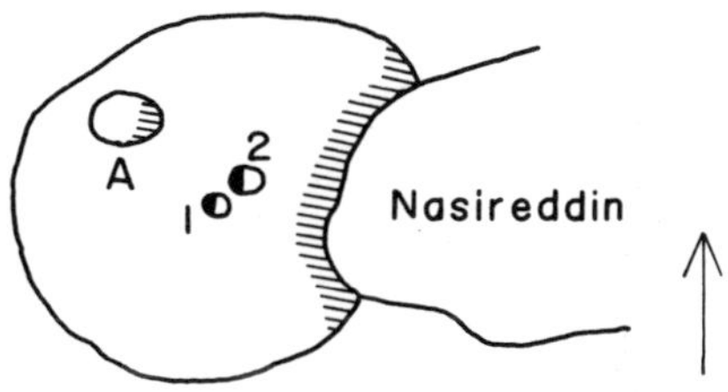

41. Huggins (-019,-656;63)

(1) ⌞6;1.0w⌟ A small central mound at the bottom of the west wall of Nasireddin.

(2) ⌞5;0.8w⌟ A hill which forms an outer part of the west wall of Nasireddin.

42. Jacobi A (+144,-852;28)

(1) ⌞5;U. L.⌟ Very faint centrally elevated floor.

43. Jacobi AB (+150,-852;47)

(1) ⌞3;0.4⌟ A pyramidal peak at the bottom of the wall of Jacobi A.

44. Jacobi C (+092,-864;35)

(1) ⌞5;0.2⌟ Very faint round mound.

45. Jacobi D (+090,-873;21)

(1) ⌞7/3;U. L.⌟ An oblong hill on the slope of the south wall.

46. Jacobi F (+087,-853;42)

(1) ⌞4;0.1⌟ A round mound at the bottom of the east wall.

47. Kinau (+127,-873;42)

(1) ⌞8;0.7e⌟ Large central mountain.

(2) ⌞2;0.2⌟ (3) ⌞2;0.1⌟ Small mounds on the east floor.

Kinau

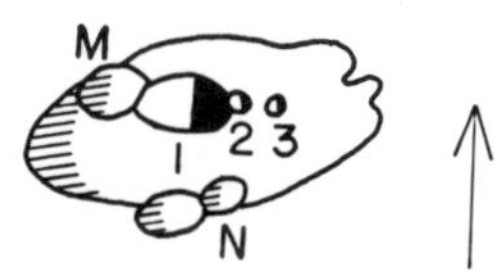

48. Kinau D (+156,-871;27)

(1) ⌞2;U. L.⌟ A faint mound.

49. Lexell B (-046,-605;23)

(1) ⌞4;0.6⌟ A central peak.

Licetus

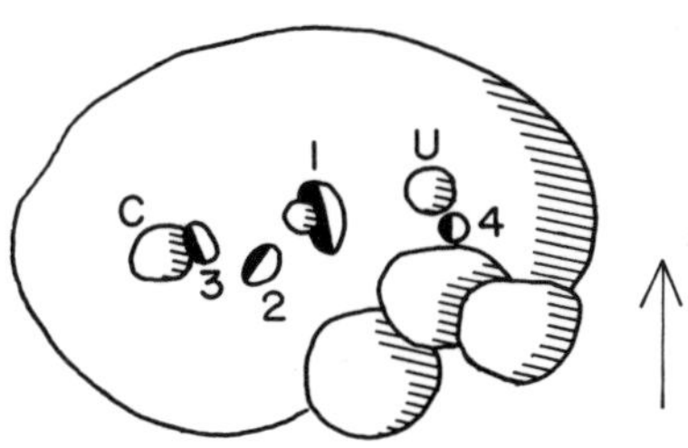

50. Licetus (+079,-733;75)

(1) ⌞10/5;0.7⌟ an oblong mountain, having a shallow crater on its west side.

(2) ⌞3;0.3⌟ (3) ⌞4;0.3⌟ Hills having westerly steep slope, forming walls of shallow crater.

(4) ⌞2.5;0.3⌟ Round mound.

51. Licetus, U. F. on the S. of (+084,-744;24)

(1) ⌞3;U. L.⌟ Very faint central peak.

52. Licetus E (+024,-704;25/18)

(1) ⌞6/3;U. L.⌟ An oblong low hill.

(2) ⌞4/3;U. L.⌟ Narrow uncinate hill, which is probably edge of an unnamed crater.

Licetus E

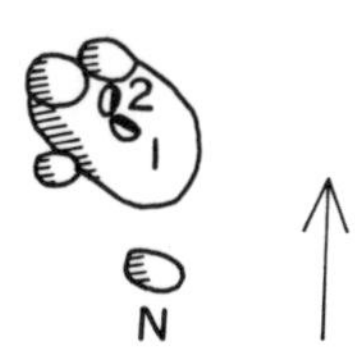

53. Lilius (+063,-814;61)

(1):α ⌞12;1.4⌟ A large central peak.

54. Lilius A, U. F. on the S. of (+081,-829;15)

(1) ⌞3;0.1⌟ A round mound stands near the edge of the south wall of Lilius A.

(XXIII)

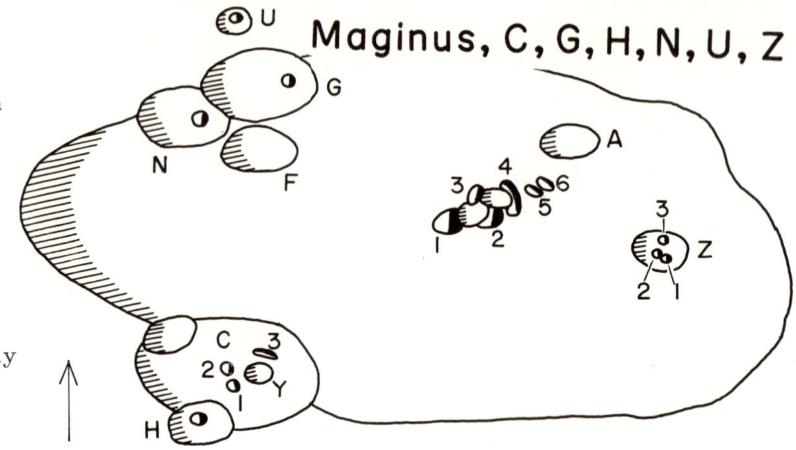

55. Lilius B (+040,-798;29)
   (1) ⌞6/3;0.2⌟ An oblong hill, whose northern end is higher, lies on the easterly eccentric position of the floor.

56. Lilius C (+034,-813;39)
   (1) ⌞3;0.2⌟ A faint mound on the westerly eccentric position of the floor.

57. Maginus (-069,-766;162)
   On the floor, there are many hills which are remnants of walls of collapsed craters. Here, only remarkable objects are registered.
   (1) ⌞10;0.5⌟ An oblong mountain.
   (2) ⌞6;0.4⌟ A hill having a small peak on its top. The (1) and (2) are separated by a shallow crater, whose diameter is 8 km.
   (3) ⌞5;0.3⌟ (4) ⌞10/2.5;0.2⌟ Ridges, forming a wall of a collapsed crater, whose diameter is 9 km.
   (5) ⌞3;0.1⌟ (6) ⌞3;0.1⌟ Small parallel hills.

58. Maginus C (-101,-785;46)
   (1) ⌞2.5;0.1⌟ (2) ⌞2;0.1⌟ Round hills.
   (3) ⌞5/2;U. L.⌟ These hills form a concentric circle with Maginus Y.

59. Maginus G (-090,-743;23)
   (1) ⌞1.5;<0.1⌟ Very tiny mound on the bottom of the east wall.

60. Maginus H (-106,-793;15)
   (1) ⌞2;U. L.⌟ Very faint central elevation.

61. Maginus H, U. F. to the W. of (-120,-791;7)
   (1) ⌞1.5;U. L.⌟ A faint mound on the slope of the east wall.

62. Maginus N (-104,-749;24)
   (1) ⌞3;<0.2⌟ A low mound on the easterly eccentric position of the floor.

63. Maginus U (-096,-736;9)
   (1) ⌞1.5;<0.1⌟ A very tiny mound.

64. Maginus Z (-040,-768;14)
   (1) ⌞2;0.1⌟ A tiny peak.
   (2) ⌞2;U. L.⌟ A white patch.
   (3) ⌞2;0.1⌟ A tiny mound.

65. Maurolycus (+180,-666;114)
   (1) ⌞24/12;2.0⌟ A large mountain, containing a small crater whose diameter is about 5 km.
   (2) ⌞10/7;1.3⌟ A hill parallel to the (1).

66. Maurolycus A (+178,-688;16)
   (1) ⌞4;0.4e/0.8w⌟ A round central mound.

(XXIII)

Maurolycus, A

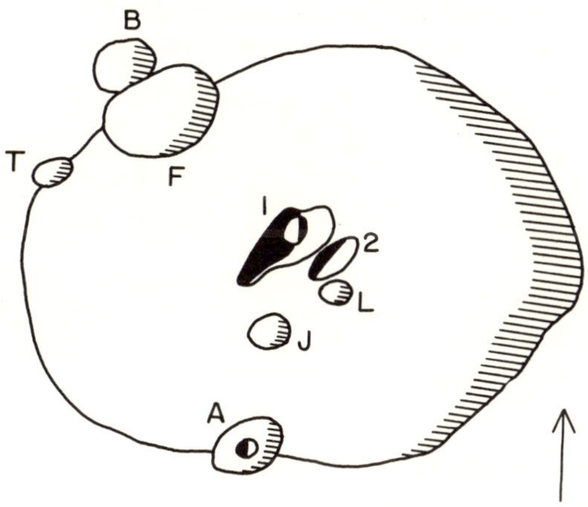

67. Maurolycus D (+177,-630; 45)
(1) [12/7; 0.3e] (2) [10/7;0.2e]
Both are rugged and uncinate hills, which form east and west part of wall of a very decrepit crater.

Maurolycus D

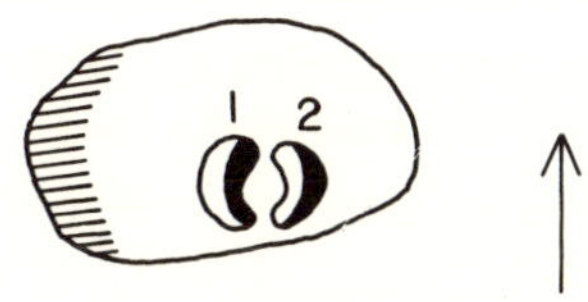

68. Miller (+010,-634;61)
(1) [7;1.4] A pyramidal mountain.
(2) [4;0.4] (3) [4;0.4] Mounds.
69. Miller A (+025,-610;39)
(1) [6;0.4] A broad hill.
(2) [1.5;0.1] A minute and oblong hill.
(3) [3;0.3] Slightly uncinate hill.
The (1) and (2) form the wall of concentric crater with Miller A. Between the (2) and (3), there is a minute crater, whose diameter is about 2 km.
70. Miller B (+014,-610;12)
(1) [2;U. L.] Very faint mound.
71. Miller C (-002,-619;40)
(1) [10;1.4w] Very broad, irregular-formed hill.
72. Montanari C, U. F. on the E. of (-191,-740 ;9)
(1) [1;U. L.] Very minute central peak.

Miller, A, B, C.

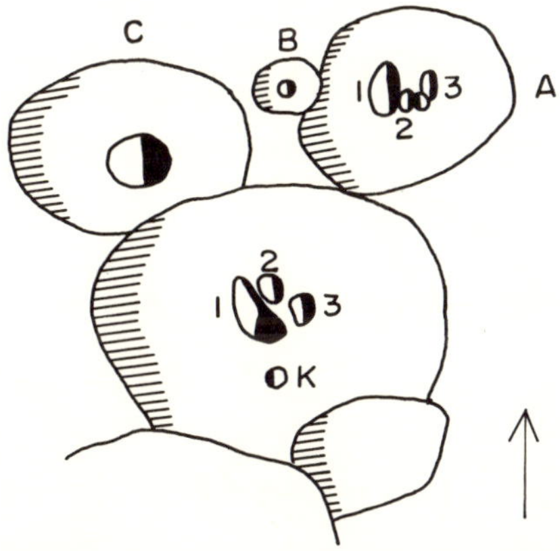

73. Moretus (-032,-943;114)
(1) [11;2.4] A large pyramidal peak.
(2) [4;0.1] A mound.
(3) [4;0.2] A mound at the bottom of the (1).
(4) [4;U. L.] A faint hill.
(5) [6;U. L.] Two-peaked hill.
(6) [13/6;U. L.] A long, irregular-formed hill.
(7) [5;U. L.] A low hill.
(8) [4;U. L.] A low mound.
74. Moretus, U. F. on the S. of (-020,-957;27)

Moretus, U. F. on the South of Moretus

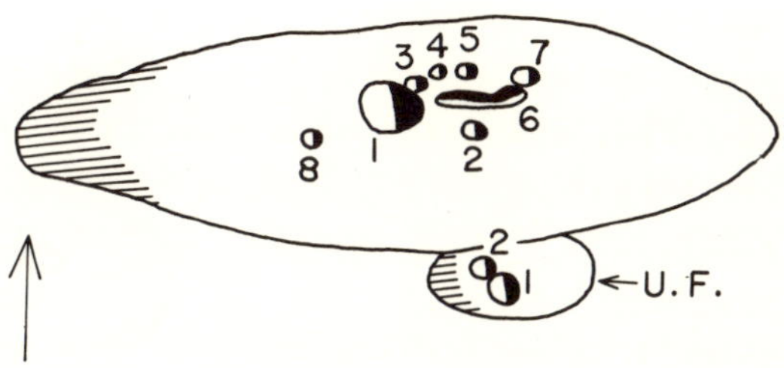

(XXIII)

(1) [5;0.4] A mound on the south wall.

(2) [4;0.2] A mound on the westerly eccentric position.

75. Mutus H (+182,-896;20)

(1) [1.5;U. L.] A minute mound.

(2) [7/4;0.3] An oblong hill which forms east wall of a small crater.

Mutus H

76. Nasireddin (+003,-657;51)

Nasireddin

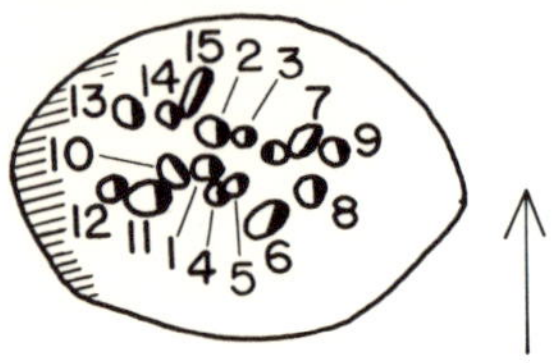

(1) [3;0.4] A tiny, but the highest peak.

(2) [3;0.2] A broad hill.

(3) [2.5;0.2] A small peak.

(4) [2;U. L.] A faint mound.

(5) [3;0.1] A low hill.

(6) [9/4;U. L.] An oblong hill.

(7) [6/4;0.2] An uncinate hill.

(8) [4;U. L.] Very faint elevation.

(9) [2;0.2] A distinct mound.

(10) [4;0.3] A low hill.

(11) [5;0.3] A low mountain.

(12) [2.5;U. L.] Faint hill.

(13) [2.5;U. L.] A faint mound.

(14) [4;0.2] A low peak.

(15) [7/1.5;0.2] A narrow dyke, whose north end forms a peak.

77. Orontius (-051,-647;109)

(1) [5/4;U. L.] A broad hill.

(2) [4/2;U. L.] A twin-peaked hill.

Orontius

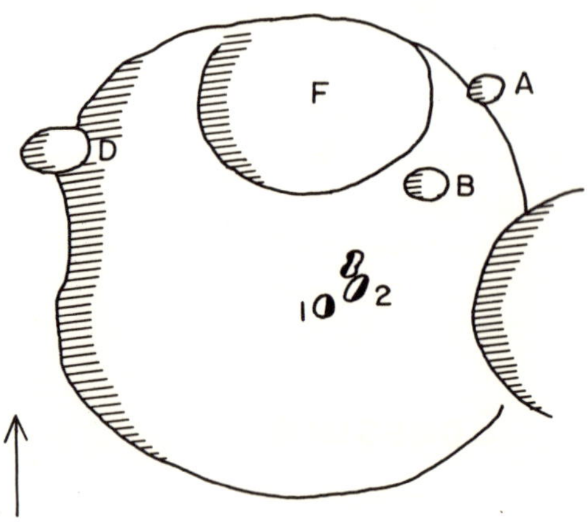

78. Pentland (+086,-903;56)

(1) [7;0.7] A peak on the westerly eccentric position of the floor.

(2) [7;U. L.] Faint elevation at the center of the floor.

79. Pentland C (+119,-906;37)

(1) [5;U. L.] A low central mound.

(2) [3;0.2] A small hill.

80. Pentland D (+110,-892;34)

(1) [10/5;U. L.] On the south part of the crater, there is a broad abd low elevation.

Pentland

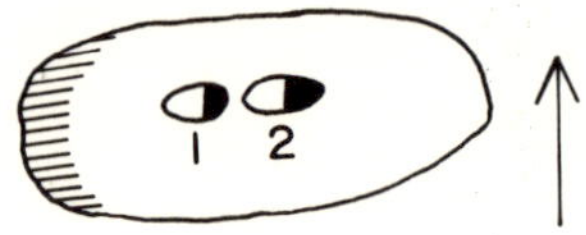

Pentland C

81. Pictet (-093,-689;62)

(1) [7/3;0.3] A low ridge.

(2) [12/2;U. L.] Faint, unsmooth dyke.

(3) [10/4;0.5] (4) [10/5;0.5] Two hills separated by a

shallow crease. The objects from (1) to (4) are all parallel to the grid structure around this formation.

(5) ⌊5;0.5w⌋ A peak having steep slope towards the west.

(6) ⌊3;0.4⌋ A small mound.

Pictet, A

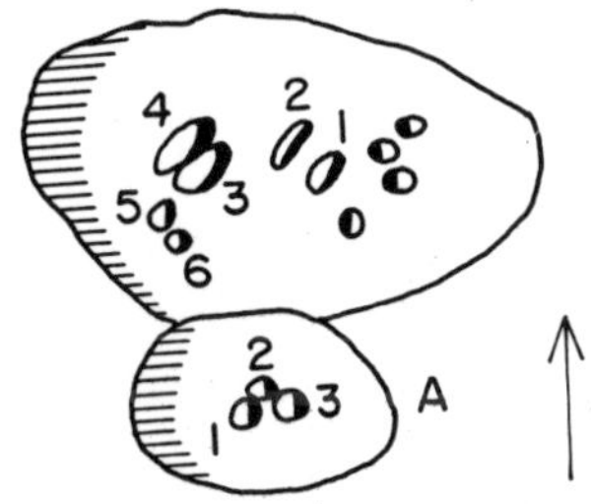

82. Pictet A (-097,-707;34)

(1) ⌊4.5;0.3⌋ The largest mound on the floor.

(2) ⌊3;0.2⌋ A mound on the center of the floor, which is probably the true central peak. The (1) and (2) form a twin peak.

(3) ⌊4;U. L.⌋ Very faint mound.

83. Pictet A, U. F. on the S. of (-094,-718;23)

(1) ⌊4;0.6⌋ Very distinct central peak, although the wall of this formation seriously damaged.

84. Porter (-098,-830;52)

This crater was formerly called Clavius B. The configuratin of this formation is shown in the figure of Clavius.

The continuation of the wall of Clavius is perceived on the north part of the floor.

(1) ⌊6;1.1⌋ A peaked mountain.

(2) ⌊6;0.5e/1.2w⌋ Peak attached to the (1).

85. Proctor H, U. F. to the E. of (-017,-720;11)

(1) ⌊4;0.4⌋ A round mound.

86. Rutherfurd (-102,-874;54/48)

The configuration of this formation is shown in the figure of Clavius.

(1) ⌊10;1.4⌋ A large pyramidal peak.

(2) ⌊4;0.4⌋ A low peak.

(3) ⌊1.5;0.2⌋ (4) ⌊2.5;0.5⌋ (5) ⌊1.0;0.1⌋ (6) ⌊3;0.3⌋ (7) ⌊1.5;0.2⌋ Group of small peaks on the east bottom of the (1). The floor is very rugged, forming a contrast to that of Porter (Clavius B).

Saussure

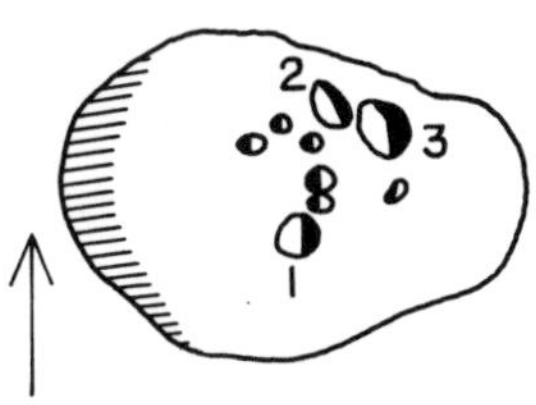

87. Saussure (-048,-686;56)

(1) ⌊3.5;0.2⌋ A round mound on the south part of the floor.

(2) ⌊7/3;U. L.⌋ A low ridge.

(3) ⌊8;U. L.⌋ A faint unsmooth elevation of triangular form.

88. Saussure CA (-006,-710;16)

(1) ⌊2;0.1⌋ A faint central mound.

Schomberger

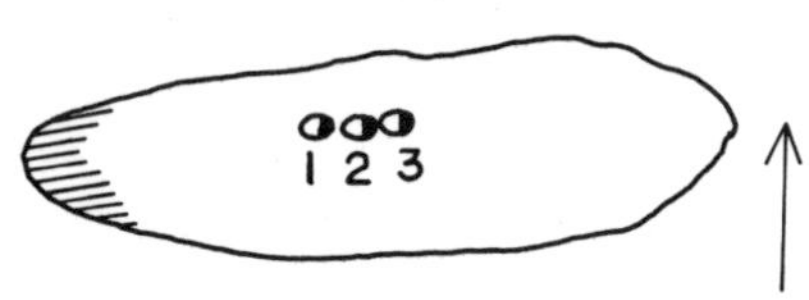

89. Schomberger (+097,-973;85)

(1) ⌊2;0.2w⌋ A small peak.

(2) ⌊4;0.25⌋ The main peak.

(3) ⌊3;0.25e⌋ Another peak on the east of the

U.F. on the W. of Schomberger C

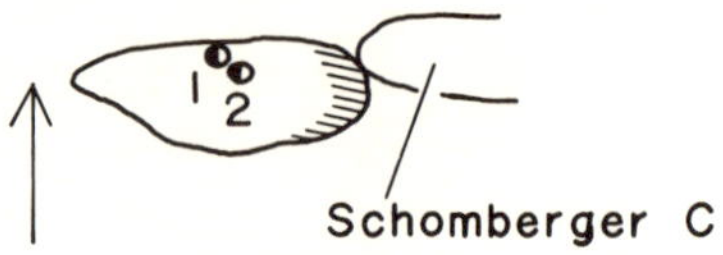

(XXIII)

(1).

90. Schomberger, U. F. to the N. of (+095,-964;37)
    (1) ⌜8/4;U. L.⌟ A dyke lies on the slope of the east wall.
91. Schomberger C, U. F. on the W. of (+042,-975;21)
    (1) ⌜3;U. L.⌟ A faint central mound.
    (2) ⌜2;U. L.⌟ A minute mound on the bottom of east wall.
92. Schomberger D (+118,-959;24)
    (1) ⌜5̇;U. L.⌟ There is a faint elevation on the bottom of the east wall.
93. Scott (+100,-990;108)
    (1) ⌜18;U. L.⌟ A central mountain having irregular form, according to the photographs of Lunar Orbiter IV.
94. Short (-034,-964;71)
    (1) ⌜7;0.6⌟ (2) ⌜5̇;0.4⌟ The (1) is a distinct peak. The (2) is smaller peak. These objects form wall of a shallow crater, whose diameter is 7 km.

Short
(Based on the Lunar Orbiter IV Photographs)

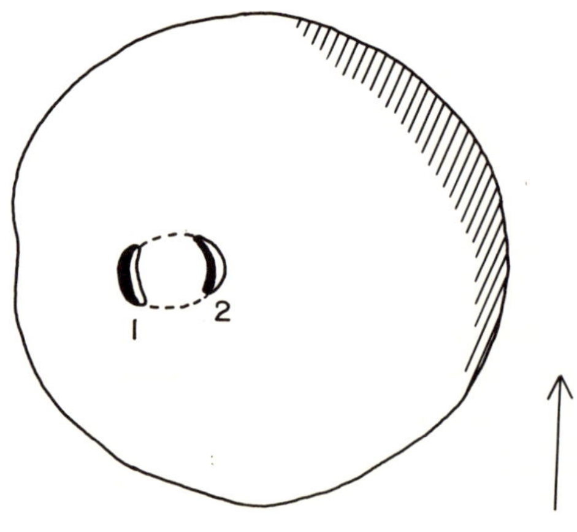

95. Simpelius (+077,-956;70)
    (1) ⌜5;0.2⌟ Very low central mountain having a summit pit.
96. Simpelius A (+097,-940;60)
    (1) ⌜4;0.2⌟ A small central peak having irregular form, according to the Photographs of the Lunar Orbiter IV.
97. Simpelius D (+047,-949;54)
    (1) ⌜4;U. L.⌟ Very faint central mound.
98. Stöfler E (+073,-692;16)
    (1) ⌜3/1;U. L.⌟ Very faint ridge.
99. Stöfler G (+025,-687;21)
    (1) ⌜6;U. L.⌟ Centrally elevated floor.
100. Stöfler H (+022,-648;24)
    (1) ⌜5.5;0.5̇⌟ Very low mountain.
101. Stöfler J (+034,-670;73)
    (1) ⌜10;1.2w⌟ A large central mountain stands on the outer slope of the west wall of Stöfler.
102. Stöfler L (+107,-631;17)
    (1) ⌜5.5/1.5;U. L.⌟ A faint oblong hill on the floor. A dyke, whose length is 20 km, runs through the south wall.
103. Stöfler P (+094,-687;35/27)

Stöfler L

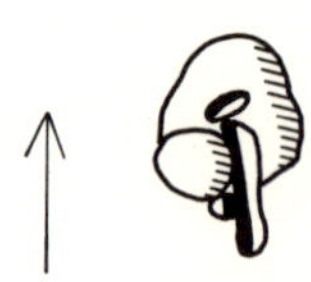

Stöfler P

(1) ⌞4;0.8w⌟ This peak is west wall of a summit crater, whose diameter is about 4 km.

(2) ⌞7/4;0.3⌟ A ridge.

(3) ⌞6;0.7⌟ An oblong ridge.

(4) ⌞5;0.5⌟ A ridge.

104. <u>Street H</u> (-140,-747;29)

(1) ⌞1.5;0.1⌟ Very minute central mound.

105. <u>Tannerus, U. F. to the S. of</u> (+192,-840;8)

(1) ⌞1;U. L.⌟ Very minute central peak.

106. <u>Tannerus E</u> (+188,-830;25)

(1) ⌞8/4;0.2⌟ An oblong hill lies on the bottom of the south wall.

107. <u>Tycho</u> (-141,-685;85)

(1) ⌞13;2.0⌟ A pyramidal peak.

(2) ⌞4;0.6⌟ Low mountain.

(3) ⌞9/3;0.2⌟ Low hill.

(4) ⌞1.5;0.1⌟ (5) ⌞2.0;0.1⌟ Minute mounds.

(6) ⌞4/2;0.1⌟ A low hill.

(7) ⌞1.5;0.2⌟ Minute peak.

(8) ⌞2;<0.1⌟ A faint mound.

(9) ⌞1.5;<0.1⌟ Minute peak.

(10) ⌞8/4;<0.2⌟ Broad and rugged elevation.

Tycho

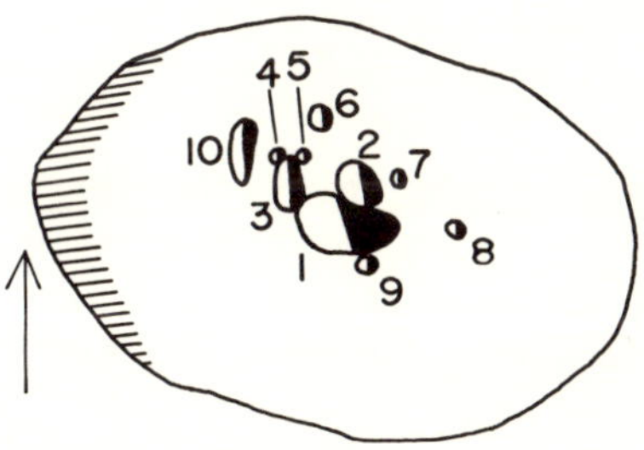

108. <u>Tycho A</u> (-159,-640;31/28)

(1) ⌞7;0.9⌟ A typical central peak.

109. <u>Tycho F</u> (-171,-655;16)

(1) ⌞2.5;U. L.⌟ A faint mound on the northerly eccentric position of the floor.

110. <u>Tycho U</u> (-179,-657;17)

(1) ⌞3;U. L.⌟ A faint, short ridge on the center.

111. <u>Zach</u> (+045,-874;71)

(1) ⌞10;0.6⌟ A twin-peaked mountain on the northerly eccentric position of the floor.

112. <u>Zach D</u> (-065,-883;29)

On the west part of the floor, there are two floor craters.

(1) ⌞6;0.4w⌟ A broad elevation.

113. <u>Zach F</u> (+028,-866;18)

(1) ⌞6/4;0.4e⌟ A ridge, which is a part of wall of a superposed crater.

XXIV

1. Asclepi (+246,-820;42)
   (1) ⌞3;<0.1⌟ Low central mound.
2. Baco (+206,-777;70)
   (1) ⌞2;U. L.⌟ Very faint mound on the westerly eccentric position of the floor.
3. Baco A (+209,-797;39)
   (1) ⌞4;0.4⌟ A typical central mound.
4. Barocius (+205,-706;82)
   (1) ⌞5̇;0.5̇⌟ A central mound.
   (2) ⌞6;0.4⌟ A hill. The (1) and (2) form a wall of shallow, small crater, whose diameter is about 5 km.
   (3) ⌞5̇;0.4⌟ A broad hill.
   (4) ⌞6;0.5̇⌟ (5) ⌞5̇;U. L.⌟ Low hills which are parts of an obscure crater, whose diameter is about 16 km.

Barocius

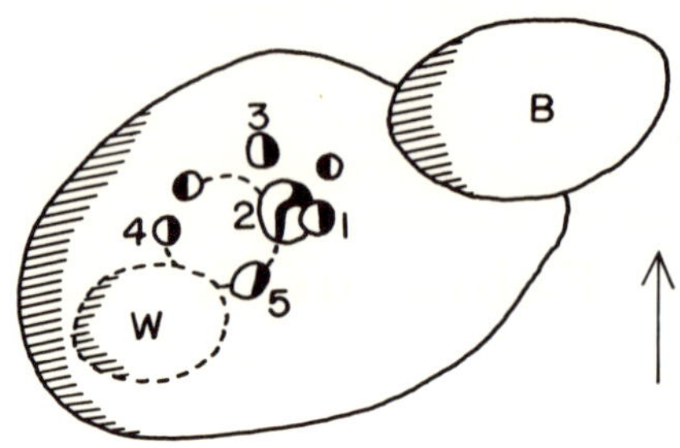

5. Barocius M (+247,-674;17)
   (1) ⌞3;U. L.⌟ Very faint mound on the center.
6. Barocius M, U. F. on the S. of (+242,-680;14)
   (1) ⌞5/3;U. L.⌟ Very faint ridge runs from the south-west wall of Barocius M.
7. Biela (+449,-818;76)
   (1) ⌞20/8;1.0̇⌟ The central mountain is a ridge, which is parallel to the grid system around this formation. On its north end, there is a shallow crater, whose diameter is about 5 km.
8. Biela A (+484,-797;26)
   (1) ⌞2;U. L.⌟ Very faint central mound.
9. Biela B (+420,-834;43)
   (1) ⌞4;U. L.⌟ Very faint mound near the north-west wall.
10. Biela C (+470,-811;26)
   The wall of Biela runs through this formation.
   (1) ⌞6;U. L.⌟ Centrally elevated floor.
   (2) ⌞4;0.4⌟ A mound on the slope of the south wall.

Biela C

11. Boguslawsky D (+218,-956;24)
   (1) ⌞2;U. L.⌟ A mound is seen on the west part of the floor.
12. Boguslawsky F (+202,-967;30)
   (1) ⌞5;U. L.⌟ A mound is seen on the east part of the floor.
13. Boussingault C (+314,-907;24)
   (1) ⌞6;U. L.⌟ A low and broad mound.
14. Boussingault E (+282,-922;98)
   (1) ⌞5̇;0.9⌟ A broad mound.
   (2) ⌞4;0.5̇⌟ A mound on the slope of the south wall.
15. Boussingault K (+279,-933;29)

(1) ⌊8;U. L.⌋ Centrally elevated floor.

16. Brenner A (+490,-648;31)

(1) ⌊5̇;U. L.⌋ Very low mound.

17. Buch A (+228,-656;22)

(1) ⌊4;U. L.⌋ Very faint mound.

18. Büsching (+269,-616;52)

(1) ⌊3;U. L.⌋ A faint mound between two shallow craters.

Boussingault C, E

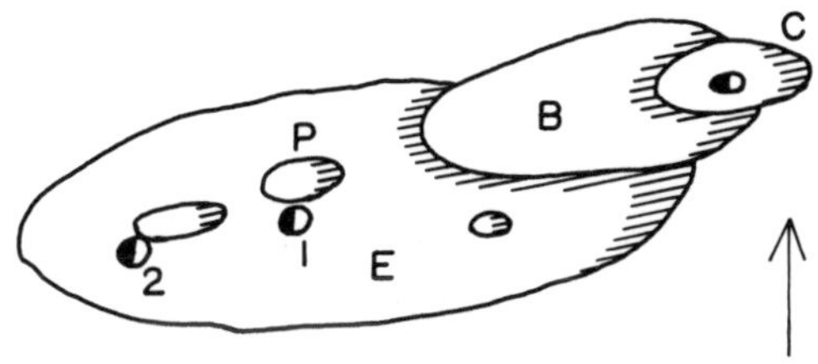

19. Fabricius (+490,-680;78)

There is concentric wall of the floor.

(1) ⌊13/8;1.4⌋ A peaked ridge, whose northern part is highest. The ridge is parallel to the linear structure on the floor of Janssen.

Fabricius A

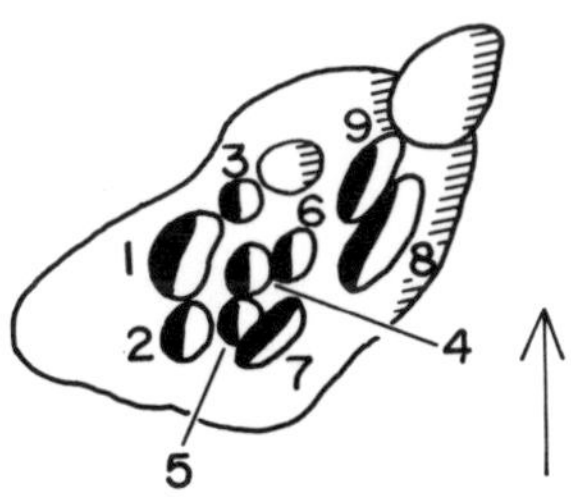

20. Fabricius A (+495,-702;45)

The floor is rich in gentle hills.

(1) ⌊15/10;0.5w/1.5e⌋ An oblong elevation.

(2) ⌊7;1.6w⌋ Distinct mound.

(3) ⌊6;0.8w⌋ Broad mound.

(4) ⌊6;1.4w⌋ A mound, standing on the lowest part of the floor, and it may be the true central peak.

(5) ⌊6;1.2w⌋ (6) ⌊6;1.0w⌋ Mounds.

(7) ⌊14/5̇;1.5w⌋ (8) ⌊15/5;0.8w⌋ (9) ⌊10/5;1.5⌋ Ridges which are collapsed wall.

21. Hagecius (+366,-864;76)

(1) ⌊5̇;0.5̇⌋ Round mound.

(2) ⌊15/4;0.3⌋ A ridge on the south-east part of the floor.

(3) ⌊3;0.5w⌋ East part of wall of a small crater.

(4) ⌊3;0.2w⌋ West part of a small crater.

Hagecius

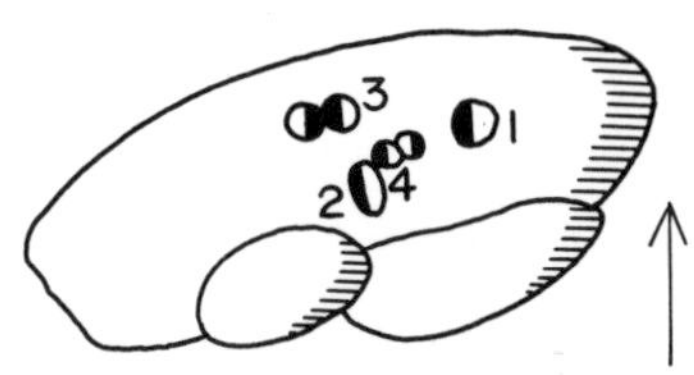

22. Hagecius C (+361,-872;24)

(1) ⌊5;1.0⌋ A ridge runs on the easterly eccentric position. This formation is a compound crater, and the ridge is the south-west wall of the smaller component.

Hagecius E

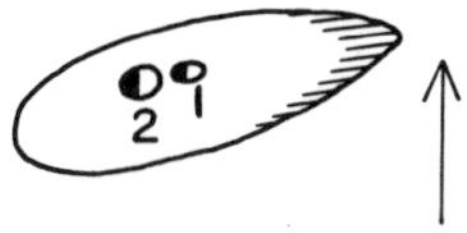

23. Hagecius E (+340,-893;44)

(1) ⌊3;U. L.⌋ Very faint mound, which may be the true central peak.

(2) ⌊5;U. L.⌋ A larger mound to the south-west of the (1).

24. Hagecius F (+361,-872;24)

(1) ⌊4;U. L.⌋ Very faint mound on the northerly eccentric part of the floor.

25. Hagecius K (+380,-876;31)

(1) ⌊4;U. L.⌋ A faint mound on the center of this formation.

26. Helmholz (+335,-916;95)

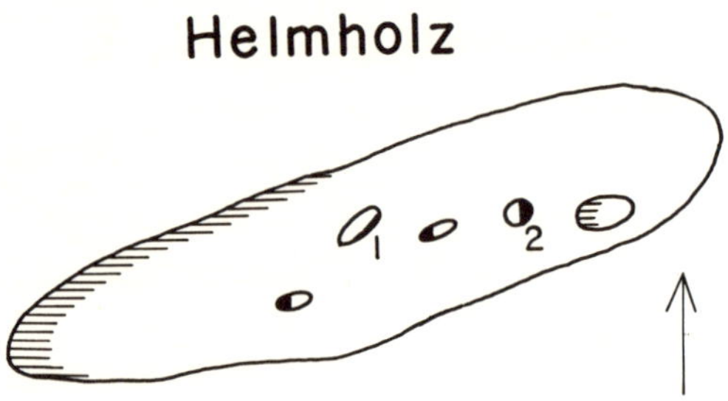

(XXIV)

(1) ⌜15/5;U. L.⌝ An oblong hill on the westerly eccentric position if the floor.

(2) ⌜6;0.5⌝ A mound on the bottom of the south-east wall.

27. Helmholz D (+326,-916;46)
There is a floor crater on the center, of which diameter is about 8 km.

(1) ⌜5.5;0.6⌝ A distinct mound on the northerly eccentric position of the floor.

28. Hommel (+315,-815;125)

(1) ⌜18;1.5⌝ A large central mountain, on whose north slope, lies a crater.

29. Hommel A (+334,-806;51)

(1) ⌜6;1⌝ A central peak superposing on the east wall of Hommel V.

30. Hommel C (+285,-817;53)

(1) ⌜7;U. L.⌝ (2) ⌜5;U. L.⌝ Very low hills.

31. Idler L (+262,-757;36) A

(1) ⌜5;0.5⌝ A distinct central peak stands on the outer bottom of the east wall of Idler.

32. Janssen (+469,-706;189)
The north border of this formation is obscure. Here, the author regards the ridge between Lockyer and

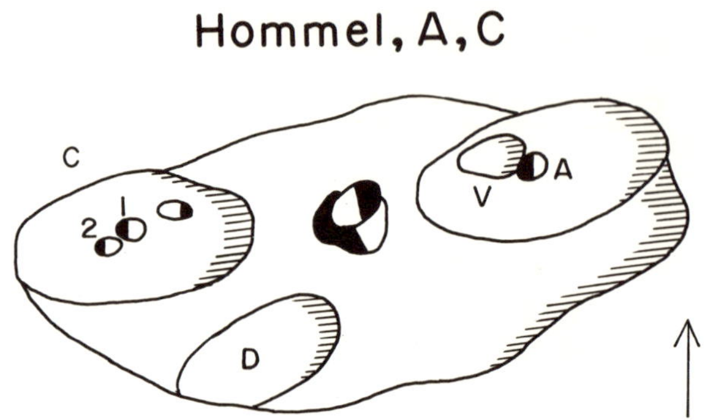

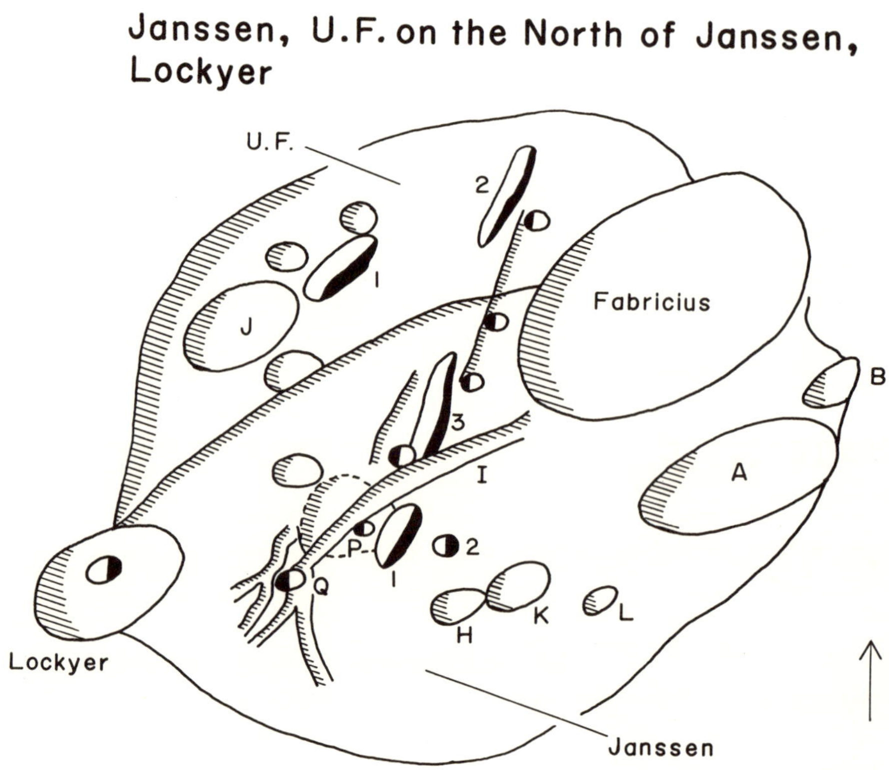

Fabricius as the north border. Hence, the crescent-formed region containing Janssen J is regarded as another formation. The floor is very rugged, accordingly only distinct objects are registered.

(1) ⌞28/13;1.7⌟ A large ridge.

(2) ⌞10;0.7⌟ A mountain which may be the true central peak.

(3) ⌞35/8;1.2⌟ A long ridge forming east bank of a groove.

33. Janssen, U. F. on the N. of (+460,-683;158)

A crescent-formed formation. The floor is very rugged, and only eminent objects are registered.

(1) ⌞30/15;1⌟ A mountain.

(2) ⌞35/10;1⌟ Long ridge, parallel to the grid structure around this formation.

34. Janssen E (+423,-752;25)

(1) ⌞8;U. L.⌟ Very faintly centrally elevated floor.

35. Janssen J (+433,-687;30)

(1) ⌞9;U. L.⌟ Very faintly centrally elevated floor.

36. Janssen X (+402,-680;24)

(1) ⌞5;0.6⌟ A peak.

(2) ⌞4;0.5⌟ A mound on the west part of the floor.

Janssen X

37. Lockyer (+413,-722;34)

(1) ⌞8;0.8⌟ A large mound on the westerly eccentric position of the floor.

38. Lockyer F (+402,-737;20)

(1) ⌞8/4;U. L.⌟ An oblong, faint hill of the westerly eccentric position of the floor.

39. Lockyer G (+385,-715;24)

(1) ⌞5.5;U. L.⌟ A large central mound.

40. Lockyer G, U. F. to the S. of (+385,-724;15)

(1) ⌞4;U. L.⌟ A faint central elevation.

41. Mallet B (+541,-727;33)

(1) ⌞8;U. L.⌟ A faint central mound.

42. Mallet D (+582,-720;42)

(1) ⌞7;U. L.⌟ (2) ⌞6;U. L.⌟ Faint mounds, forming east bank of Vallis Rheita.

Mallet D

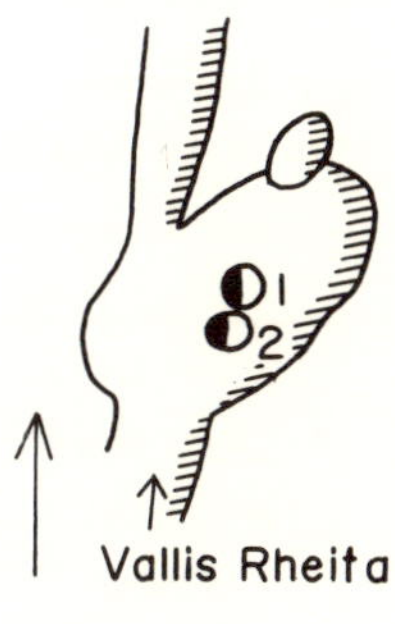

43. Mallet J (+547,-751;52)

(1) ⌞10/6;0.6e⌟ A ridge, which is remnant of wall of a collapsed crater.

(2) ⌞4;U. L.⌟ A faint hill.

(3) ⌞4;0.7⌟ A mound on the north part of the floor.

44. Metius (+523,-647;88)

(1) ⌞13/4;0.8⌟ A ridge.

(2) ⌞4;0.6⌟ A mound.

(3) ⌞5;0.8⌟ A peak.

Metius

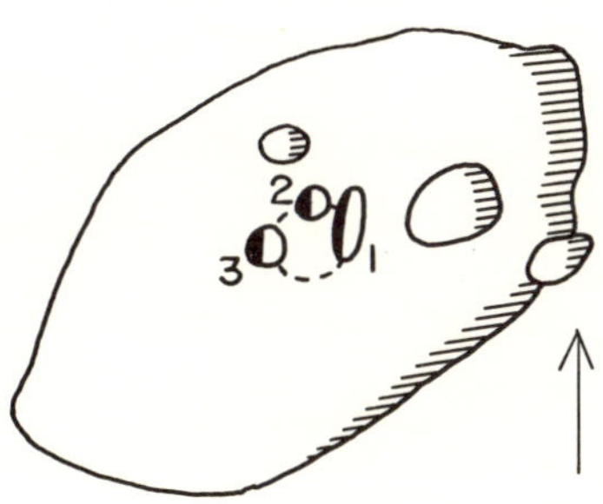

Mallet J

These three objects forms a wall of collapsed crater whose diameter 13 km.

45. Metius, U. F. to the N. of (+499,-621;23)

(1) ⌊10;U. L.⌋ Very faintly centrally elevated floor.

46. Mutus V (+237,-890;24)

(1) ⌊5;U. L.⌋ Broad mound at the center of the floor.

47. Mutus W (+255,-918;21)

(1) ⌊2;U. L.⌋ A tiny mound on the center of the floor.

48. Nearch A (+321,-867;43)

(1) ⌊5̇;0.5⌋ A mountain on the westerly eccentric position of the floor.

(2) ⌊2;U. L.⌋ Very faint mound, which may be the true central peak.

Nearch A

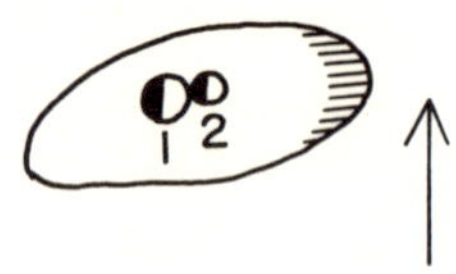

49. Nearch B (+284,-874;43)

(1) ⌊4;U. L.⌋ A central peak, superposing on the wall of Nearch BA and BC.

Nearch B

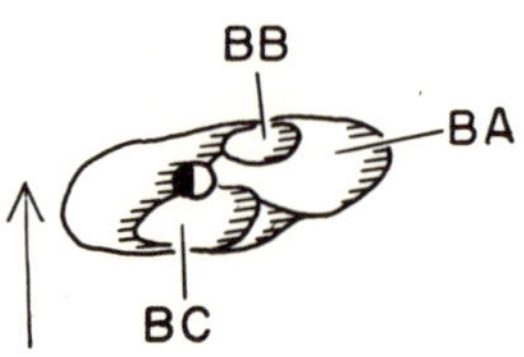

50. Nicolai Z, U. F. to the W. of (+268,-659;9)

(1) ⌊2;U. L.⌋ A small peak stands on the slope of the north wall.

51. Nicolai Z, U. F. to the N.-W. of (+267,-645;11)

(1) ⌊2.5;U. L.⌋ A small mound stands on the slope of the north-west wall.

52. Pitiscus (+324,-772;82)

(1) ⌊13;1.2⌋ Alarge central mountain, on whose north bottom lies a floor crater, named Pitiscus A.

(2) ⌊4;0.3⌋ A mound.

(3) ⌊20/6;0.4⌋ A ridge of semicircular form around the (2).

(4) ⌊6;U. L.⌋ An elevated part on the slope of the south wall.

Pitiscus

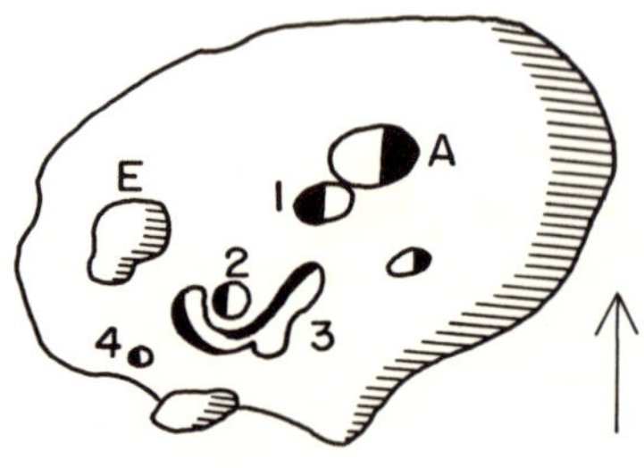

53. Pontécoulant (+475,-854;91)

(1) ⌊3;U. L.⌋ Faint mound, forming the west wall of Pontécoulant H.

(2) ⌊5̇;U. L.⌋ A larger mound on the westerly eccentric position of the floor.

Pontécoulant

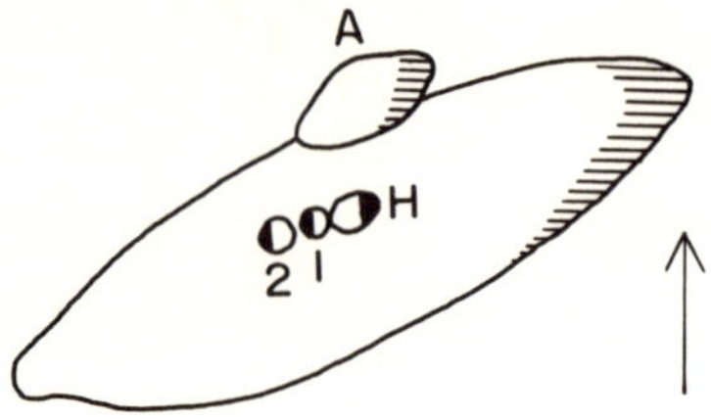

54. Pontécoulant E (+445,-870;44)

(1) ⌊12;U. L.⌋ Very broad elevation on the south-east wall.

55. Pontécoulant J (+428,-880;39)

(1) ⌊7;0.8⌋ A distinct mound on the bottom of the south wall.

56. Rheita (+585,-603;70)

(1) ⌊10.0;1.5̇⌋ A distinct peak on the northerly eccentric position of the floor.

(XXIV)

57. Riccius E (+342,-642;22)

(1) ⌜6;U. L.⌟ A faint elevation on the westerly eccentric position of the floor. This formation is a compound crater, and the elevation is trace of east wall of the smaller component.

58. Rosenberger (+388,-823;96)

(1) ⌜6;0.8⌟ A central peak standing on the south of a small deep crater.

(2) ⌜6;0.4⌟ Broad elevation at the bottom of the west wall.

(3) ⌜5;0.4⌟ A mound near the west wall.

Rosenberger

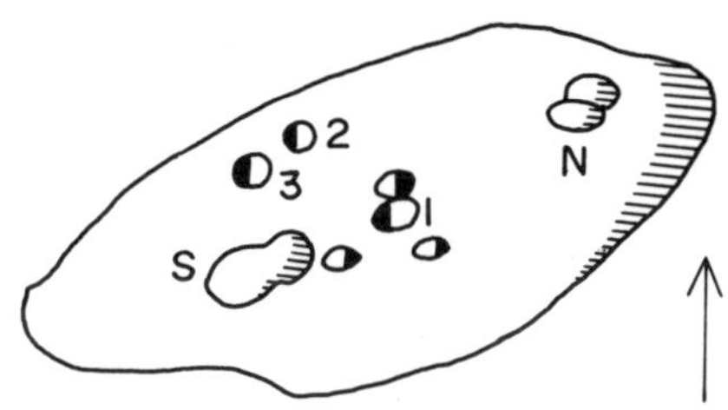

59. Rosenberger C (+412,-789;47)

(1) ⌜7;0.7⌟ A distinct central peak.

60. Rosenberger D (+366,-843;50)

(1) ⌜5;0.6⌟ A central peak.

61. Steinheil X (+484,-738;17)

(1) ⌜7;U. L.⌟ Centrally elevated floor.

62. Tannerus C (+220,-822;16)

(1) ⌜2.5;0.1⌟ A tiny central peak.

63. Tannerus H (+226,-811;20)

(1) ⌜6;U. L.⌟ Centrally elevated floor.

64. Tannerus P (+211,-825;18)

(1) (1) ⌜4;0.3⌟ A mound on the northerly eccentric position of the floor.

Vlacq

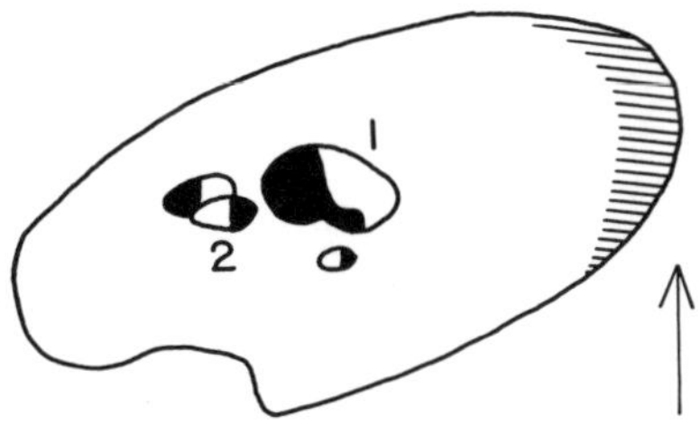

65. Vlacq (+374,-802;89)

(1) ⌜15;1.3⌟ A very large central mountain, having two peaks. Height of the lower peak is about 0.4 km.

(2) ⌜5;0.3⌟ A hill forming north wall of a small crater, whose diameter is about 10 km.

66. Vlacq G (+355,-818;27)

(1) ⌜4;U. L.⌟ A faint mound is seen on the slope of the south wall.

67. Watt (+487,-761;66)

(1) ⌜7;U. L.⌟ (2) ⌜6;U. L.⌟ (3) ⌜5;U. L.⌟ Very faint hills.

(4) ⌜35/7;U. L.⌟ Very long ridge along its west wall.

Watt

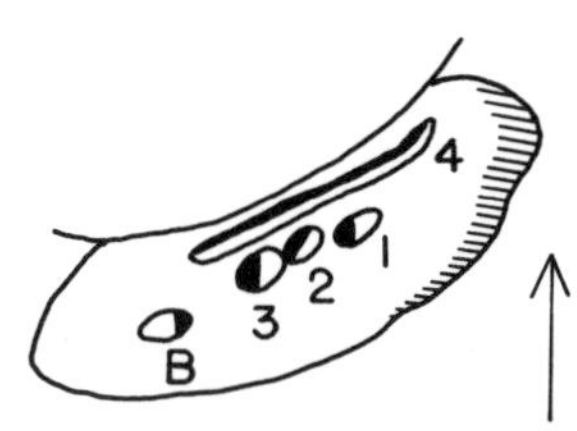

68. Watt M (+519,-800;42)

(1) ⌜10;1⌟ Very broad mound.

69. Young D (+570,-688;46)

(1) ⌜4;1.0⌟ A peak on somewhat northerly eccentric position of the floor.

XXV

1. Brisbane (+609,-756;45)
   (1) [4;U. L.] Very faint central mound.
2. Fraunhofer (+662,-636;57)
   (1) [8;U. L.] A faint hill on the south part of the floor.
   (2) [5̇;U. L.] Round mound at the bottom of the east wall.
3. Fraunhofer C (+662,-681;38)
   (1) [5̇;0.8] A distinct central peak.
4. Furnerius D (+661,-602;16)
   (1) [1.5;U. L.] Very tiny central mound.
5. Furnerius H (+742,-610;44)
   (1) [6;U. L.] A low central mound.
6. Hamilton (+731,-679;44)
   (1) [7;U. L.] A mound on the northerly eccentric position of the floor.
   (2) [3;U. L.] Smaller mound to the south of the (1).
7. Peirescius (+637,-725;62)
   (1) [15;U. L.] Very broad elevation on the northerly eccentric position.
8. Vega B (+619,-722;30)
   (1) [5̇;1.2] A central peak.

Fraunhofer

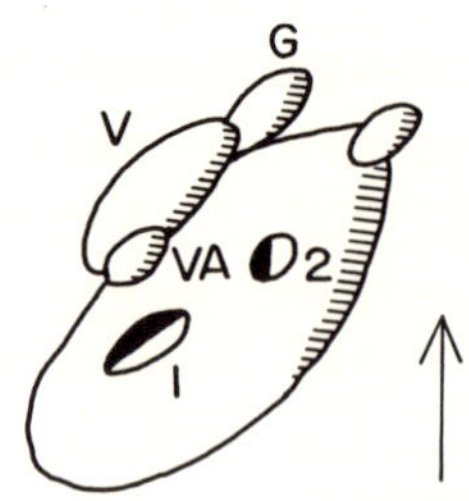

Hamilton

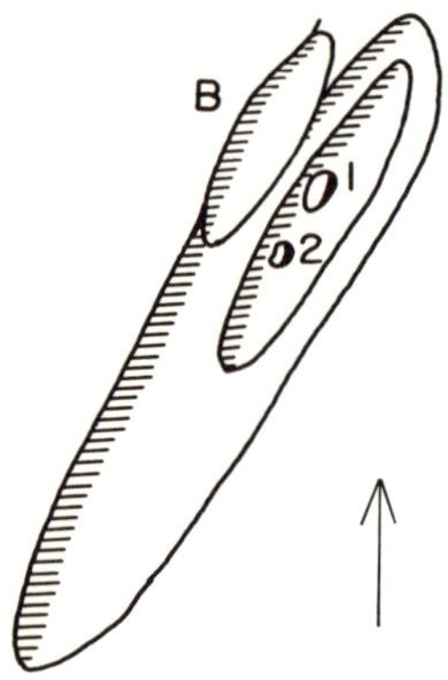

Peirescius

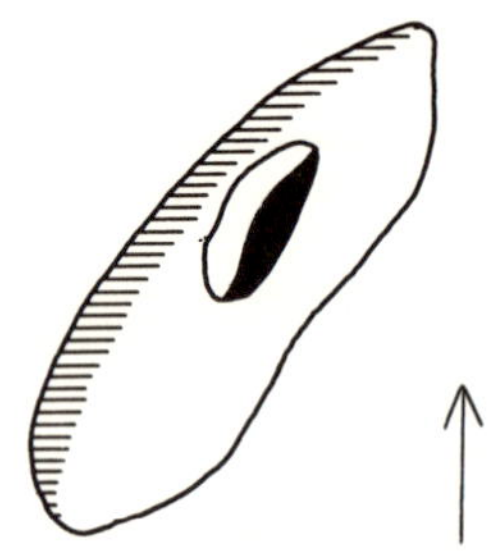

INDEX

XVIII-1	Aristarchus	-676	+402	45	ml;rc
IV-1	Aristillus	+018	+557	55	ml,p7;rc
XV-4	Aristoteles	+191	+768	87	dl,el,h2,Ml,ml,pl,ptl,r6;ps
XIV-1	Arnold A	+232	+932	57	pl
XIV-2	Arnold K	+222	+944	29	ml
VIII-18	Arzachel	-031	-313	97	h4,M3,m7,P2,pl
XXIV-1	Asclepi	+246	-820	42	ml
XIV-3	Atlas	+481	+727	87	m5,pl;tcp
XIV-4	Atlas E	+447	+750	58	pl
IV-2	Autolycus	+022	+557	39	pl,r4;tcp
XI-2	Auzout	+885	+178	33	el,hl;tcp
VIII-19	Azophi A	+178	-413	29	Ml
VIII-20	Azophi B	+170	-400	19	pl
VIII-21	Azophi G	+195	-405	53	Ml

B

XXIV-2	Baco	+206	-777	70	ml
XXIV-3	Baco A	+209	-797	39	ml
XXIII-2	Baco K	+178	-809	29	ml
XXII-1	Bailly, U. F. to the N.-W. of	-491	-869	62	ml
XVIII-2	Balboa A	-944	+299	47	pl
VIII-22	Ball	-118	-587	40	Ml
VIII-23	Ball A	-133	-570	29	m3
XXIII-3	Ball C	-120	-611	31	Ml
XXIV-4	Barocius	+205	-706	82	h4,ml
XXIV-5	Barocius M	+247	-674	17	ml
XXIV-6	Barocius M, U. F. on the S. of	+242	-680	14	rl
IX-5	Beaumont	+458	-310	53	e5,h5,M7,m2,pl,rl;ps
IX-6	Beaumont, U. F. to the S.-E. of	+471	-330	12	Ml
IX-7	Beaumont A	+447	-280	12	Ml
IX-8	Beaumont B, U. F. on the W. of	+417	-319	15	Ml
X-1	Behaim	+943	-285	56	pl
X-2	Behaim B	+935	-278	24	pl
XIV-5	Belkovich	+479	+877	198	p3;tcp
XXII-2	Bettinus	-316	-895	71	pl
XII-1	Bernouilli	+714	+574	47	pl
XVI-1	Bianchini	-372	+752	39	pl
XXIV-7	Biela	+449	-818	76	Ml;ps
XXIV-8	Biela A	+484	-797	26	ml
XXIV-9	Biela B	+420	-834	43	ml

XXIV-10	Biela C	+470	-811	26	el,ml;ps
XV-5	Birmingham	-078	+905	97	hl,p2;rf
XXIII-4	Blancanus	-163	-896	105	hl,p2
XXIII-5	Blancanus C	-187	-917	46	rl
VIII-24	Blanchinus, U. F.	+029	-434	9	ptl
	on the W. Wall of				
I-4	Bode	-042	+117	19	ml
XXIV-11	Boguslawsky D	+218	-956	24	ml
XXIV-12	Boguslawsky F	+202	-967	30	ml
XXIII-6	Boguslawsky N, U.	+161	-959	42	ml
	F. on the N. of				
X-3	Bohnenberger	+619	-279	33	e2,h2,p2
X-4	Bohnenberger C	+624	-318	16	ml
III-1	Bond, G.	+500	+536	20	el
XV-6	Bond, W. C.	+026	+909	157	h8,mll,pl;rf,tcp,innumerable objects
X-5	Borda	+658	-425	44	hl,Ml
X-6	Borda, U. F.	+662	-428	23	hl
	Contained in				
I-5	Boscovich P	+175	+199	67/17	ml;curious form
XXIV-13	Boussingault C	+314	-907	24	ml
XXIV-14	Boussingault E	+282	-922	98	m2
XXIV-15	Boussingault K	+279	-933	29	el
XXI-1	Bouvard	-761	-640	179	hl,Ml
XX-1	Bouvard A	-806	-586	68	hl
XXI-2	Bouvard M	-741	-651	69	Ml,m2
XXIV-16	Brenner A	+490	-648	31	ml
XVIII-3	Briggs	-835	+445	39	el,hl,Ml
XVIII-4	Briggs B	-832	+471	25	pl
XXV-1	Brisbane	+609	-756	45	ml
XXII-3	Brown E	-207	-729	22	m2,pl
XXIV-17	Buch A	+228	-656	22	ml
VII-2	Bullialdus	-353	-354	59	p5
VII-3	Bullialdus B	-342	-397	21	hl
XII-2	Burckhardt	+714	+517	57	hl,Ml
XIV-6	Bürg	+334	+708	40	hl,Ml,ml
VIII-25	Burnham	+124	-240	22	ml,r6;ps
XXIV-18	Büsching	+269	-616	52	ml

C

XV-7	Calippus	+145	+629	31	el,m6
XV-8	Calippus A	+110	+602	16	pl
VII-4	Campanus	-411	-469	48	h2
II-6	Capella	+567	-133	45	hl
II-7	Capella F	+572	-160	13	dl

II-8	Capella F, U. F.	+575 -168	15	el
	on the S. of			
VII-4	Capuanus	-411 -469	48	h2
XVI-2	Carpenter	-273 +936	60	M2
XXIII-7	Casatus A	-184 -956	50	ml
XV-9	Cassini	+061 +646	57	h2,M2,m2,p3,rl
XV-10	Cassini A	+063 +649	17	el;cpc
IX-9	Catharina	+381 -311	97	pl
IX-10	Catharina B	+394 -292	25/22	rl
IX-11	Catharina C	+387 -348	30	pl; plateau-like floor
IX-12	Catharina P	+378 -294	49	p2
XIX-1	Cavalerius	-916 +089	64	h2;tcp
IX-13	Celsius, U. F. to	+257 -551	17	h2
	the W. of			
II-9	Censorinus C	+561 -053	30	dl,hl,Ml,m2,Pl,p2;ps,tcp
II-10	Censorinus E, U. F.	+572 -085	30	ml;cpc
	to the N. of			
II-11	Censorinus N	+596 -034	37	pl
II-12	Censorinus S	+589 -067	12	el
XIV-7	Cepheus	+543 +653	40	Ml,ml
VII-5	Cichus	-302 -549	39	h2,pl;tcp
VII-6	Cichus G, U. F. to	-332 -596	40	h2,pl;superposed crater
	the S. of			
XXIII-8	Clairaut B	+145 -747	43	r2
VII-7	Clausius C	-512 -580	15	el
VII-8	Clausius F, U. F.	-497 -595	61	cl;ccc
	around			
XXIII-9	Clavius	-130 -852	225	d4,hl,M2,r4,p2,sl
XXIII-10	Clavius D	-110 -854	28	ml
XXIII-11	Clavius K	-166 -869	20	ml;ps
XXIII-12	Clavius L	-188 -854	24	ml;ps
XII-3	Cleomedes	+730 +465	126	h2,Ml
XII-4	Cleomedes D, U. F.	+771 +485	40	ml
	Containing			
XII-5	Cleomedes E	+714 +479	21	el
X-7	Colombo	+696 -261	78	hl,p3,pt2
X-8	Colombo E, U. F.	+643 -273	23	ml
	on the W. of			
VI-1	Copernicus	-337 +168	93	el,h6,Ml,m2;tcp,rc
XXIII-13	Curtius	+030 -922	95	hl,Ml
XXIII-14	Curtius B	+036 -896	41	m2
XXIII-15	Curtius D	+060 -905	61/45	hl,pl
XXIII-16	Cuvier A	+127 -792	18	el
XXIII-17	Cuvier B	+148 -784	17	el
XXIII-18	Cuvier E	+137 -791	19	el

XXIII-19	Cuvier O	+130	-784	10	m1
XXIII-20	Cuvier Q	+114	-783	13	m1
IX-14	Cyrillus	+397	-231	93	m1,p3
IX-15	Cyrillus, U. F. to the S.-W. of	+355	-258	20	r1
IX-16	Cyrillus A	+381	-238	15	e1
IX-17	Cyrillus F, U. F. on the E. Wall of	+426	-266	22	m1
XXIII-21	Cysatus	-043	-914	49	m2

D

IX-18	Daguerre	+541	-205	45	h2,m1
XIX-2	Damoiseau	-872	-086	37	p2,pt1
VII-9	Darney	-386	-252	15	m1
XX-2	Darwin B	-895	-341	51/39	h1,r1
XI-3	da Vinci	+698	+160	31	e3,h2,m1
VIII-26	Davy	-138	-205	35	m5
VIII-27	Davy A	-131	-212	15	m2
XII-6	Debes	+687	+493	31	m1
II-13	Delambre	+300	-034	53	s2
XIV-8	de la Rue	+410	+858	136	c2,h6,m2,r3
VIII-28	Delaunay	+041	-380	50	m1,r4;ps
V-1	Delisle	-492	+500	25	m4
XII-7	Delmotte	+772	+456	33	e1,m1
XXIII-22	Deluc	-028	-819	47	d1,p1
XXIII-23	Deluc A	-004	-810	55	m1
XXIII-24	Deluc AB	+006	-823	22	d2
XXIII-25	Deluc D	-023	-832	27	e1
XXIII-26	Deluc H	-021	-810	26	e1
XXIII-27	Deluc J, U. F. on the E. of	+014	-895	7	m1
XIV-9	Democritus	+267	+886	39	m1,p1
XXIII-28	Demonax	+175	-979	114	p1
XV-11	De Sitter	+110	+984	65	d1
II-40	Dimple Crater in Mare Tranquillitatis	+4150	+0464	0.098	central bodies 3
II-41	Dimple Crater in Mare Tranquillitatis	+4158	+0474	0.031	central body 1
V-2	Diophantus	-499	+463	19	e1;cpc
II-14	Dollond B, U. F. on the S. Rampart of	+232	-145	9	M1
II-15	Dollond B, U. F. to the S. of	+231	-157	6	e1
II-16	Dollond B, U. F.	+264	-152	137	p5

	Containing				
II-17	Dollond C, U. F. to the N.-W. of	+207	-107	10	rl;ps
VIII-29	Donati	+084	-353	36	pl
VII-10	Doppelmayer	-581	-477	64	Ml,h4
XX-3	Doppelmayer G, U. F. on the S. of	-618	-492	17	ml
XXIII-29	Drygalski	-180	-982	176	Ml

E

XX-4	Eichstadt	-904	-385	49	pl
VII-11	Elger	-405	-578	21	hl
VI-2	Encke	-595	+080	29	M3
XIV-10	Endymion C	+457	+852	32	pl
XIV-11	Endymion J	+460	+804	67	ml,pl,r3
XV-12	Epigenes	-032	+923	55	h2,M2,ml,p2
XXII-4	Epimenides	-380	-655	27	el;ccc
XXII-5	Epimenides A, U. F. to the N.-E. of	-355	-679	15	rl
XXII-6	Epimenides S	-366	-664	26	el,m4
IV-3	Eratosthenes	-190	+250	58	el,Ml,r4;tcp
XIV-12	Eudoxus	+201	+698	67	h2,Ml,m5,p2,r4
V-3	Euler	-447	+395	27	Ml

F

XXIV-19	Fabricius	+490	-680	78	rl;ccc
XXIV-20	Fabricius A	+495	-702	45	el,m5,r3;tcp
XXIII-30	Faraday	+112	-675	69	cl,ml
XXIII-31	Faraday C	+102	-685	30	dl,e2,ml,p2
XXIII-32	Faraday G	+122	-718	31	Ml
XXIII-33	Faraday L, U. F. to the N.-E. of	+122	-619	8	ml
VIII-30	Faye	+062	-365	37	pl
IX-19	Fermat	+313	-385	39	pt3
IX-20	Fermat C, U. F. on the E. of	+305	-361	31	r2
IX-21	Fermat P	+304	-400	36	ml
IX-22	Fermat P, U. F. on the S. Wall of	+301	-410	8	ml
I-6	Flammarion W	-041	-037	7	Ml;ps
XIX-3	Flamsteed	-696	-078	21	hl
XX-5	Fontana	-802	-277	31	e2,m2
XV-13	Fontenelle	-145	+893	38	cl,el,hl,m4
XX-6	Fourier, U. F. to	-698	-534	78	h2,m2,pl;tcp

	the S. of				
IX-23	Fracastorius	+509	-363	124	e6,m3,p3,rl
IX-24	Fracastorius, U. F. to the N.-W. of	+505	-329	12	el
IX-25	Fracastorius A	+542	-413	18	el
IX-26	Fracastorius A, U. F. to the N. of	+533	-407	16	rl
IX-27	Fracastorius B	+560	-384	27	pl
IX-28	Fracastorius D	+477	-371	27	pl,rl;ps
IX-29	Fracastorius K, U. F. on the S. of	+521	-436	27	ml,rl
XIV-13	Franklin	+576	+627	56	ml,pl
XXV-2	Fraunhofer	+662	-636	57	hl,ml
XXV-3	Fraunhofer C	+662	-681	38	pl
X-9	Furnerius, U. F. on the W. of	+681	-584	26	el
XXV-4	Furnerius D	+661	-602	16	ml
XXV-5	Furnerius H	+742	-610	44	ml

G

XIV-14	Galle	+213	+827	21	e2,hl,ml,ptl
XVII-1	Galvani	-644	+762	75	pl
XX-7	Gassendi	-611	-268	110	dl,hll,M2,m7,p5,Pl,rl;tcp,ps
XX-8	Gassendi A	-616	-268	33	e3,h2,m3,pl
XI-4	Gaudibert	+602	-190	29	hl,Ml
X-10	Gaudibert B, U. F. to the W. of	+619	-218	20	ml;cef
IX-30	Gaudibert H	+581	-238	11	ml
XII-8	Gauss	+795	+587	177	h3,M2,p6
XIII-1	Gauss, U. F. to the N.-E. of(Riemann)	+770	+635	133	e3,pl,rl
XII-9	Gauss, U. F. to the S.-E. of(Rayleigh)	+872	+489	102	cl,Ml
V-4	Gay-Lussac A	-339	+228	14	hl
XII-10	Geminus	+689	+566	85	pl
XII-11	Geminus, U. F. on the S.	+711	+543	20	ml
VIII-31	Gemma Frisius	+191	-564	88	m5,pl,rl
IX-31	Gemma Frisius, U. F. on the N.-E. floor of	+209	-556	12	ml
VIII-32	Gemma Frisius, U. F. on the W. Wall of	+161	-570	46	Ml
VIII-33	Gemma Frisius D	+156	-564	27	el
VIII-34	Gemma Frisius EB,	+195	-600	15	el

	U. F. to the E. of				
VIII-35	Gemma Frisius EB,	+184	-594	17	M1,m1
	U. F. to the N. of				
VIII-36	Gemma Frisius H	+179	-536	28	M1,m2,p1
VIII-37	Gemma Frisius H,	+177	-525	36	h1
	U. F. on the N. of				
XXIII-34	Gemma Frisius Y, U.	+193	-612	6	m1
	F. on the S.-E. of				
XVII-2	Gerard Q	-686	-721	177	c1;ccc
XI-5	Gilbert, U. F. on	+967	-086	57	p1
	the S. of				
XI-6	Gilbert J	+952	-075	38	p1
XI-7	Goclenius	+696	-174	66/54	d3,e2,h3,p3,pt2;ps
I-7	Godin	+177	+032	35	p1;rf,rc
XV-14	Goldschmidt	-015	+957	125	m5;innumerable mounds
IX-32	Goodacre	+205	-540	46	p1
IX-33	Goodacre G	+201	-549	16	m1
IX-34	Goodacre K	+200	-514	11	p1
XIV-15	Grove	+415	+647	27	pt1
XIX-4	Grimaldi	-925	-091	222	h5,m2,r1
XXIII-35	Gruemberger	-068	-920	94	h2,M1,m1,p1,s1
VI-3	Guericke	-239	-200	58	e1,h3,m2
XI-8	Gutenberg	+652	-151	71	h3,M1,m7,r1;ps
XI-9	Gutenberg C	+649	-175	46	e2,m1,p1,r2
XI-10	Gutenberg G	+639	-106	31	p1
I-8	Gyldén	+005	-093	47	h1

H

XXIV-21	Hagecius	+366	-864	76	h2,m1,r1
XXIV-22	Hagecius C	+361	-872	24	r1;cpc
XXIV-23	Hagecius E	+340	-893	44	m2
XXIV-24	Hagecius F	+361	-872	24	m1
XXIV-25	Hagecius K	+380	-876	31	m1
XII-12	Hahn	+820	+519	84	p1
XXII-7	Haidinger	-328	-631	21	e1,m2
XXII-8	Haidinger C	-293	-630	19	m1
XXII-9	Haidinger M	-297	-608	22	p1
XXII-10	Hainzel	-415	-660	70	e1,M1,p1
XXII-11	Hainzel A	-426	-646	53	h1,M1,m1,p1
XXII-12	Hainzel B	-434	-615	15	c1;ccc
XXII-13	Hainzel C	-408	-657	38	h1,M1,m1,p1
XXII-14	Hainzel R	-463	-626	19	h1;cpc
I-9	Halley	+100	-140	36	p1
XXV-6	Hamilton	+731	-679	44	m2

XII-13	Hansen	+926	+242	40	ml
XIX-5	Hansteen	-771	-200	45	hl,Ml,ml,pl,r2
XVI-3	Harpalus	-416	+795	40	hl,m2
XIV-16	Hayn	+426	+904	87	Ml,pl
XIV-17	Hayn A	+430	+890	54	Ml
XIV-18	Hayn E	+357	+921	41	pl
XXII-15	Heinsius	-234	-636	66	ml
XXII-16	Heinsius A	-232	-636	20	el
XXII-17	Heinsius B	-245	-642	24	hl
XXII-18	Heinsius C	-233	-651	22	ml
XXIII-36	Heinsius M	-199	-656	13	ml
XXII-19	Heinsius MA	-203	-647	23	ml
XXIII-37	Heinsius P	-182	-635	37	M2,ml,pl
XXIII-38	Heinsius Q	-191	-642	37	ml
VIII-38	Hell	-114	-535	33	Ml,m5,pl,r3
VIII-39	Hell A	-121	-557	22	pl
VIII-40	Hell B	-087	-500	22	ml
VIII-41	Hell J	-104	-495	6	cl
XXIV-26	Helmholz	+335	-916	95	hl,ml
XXIV-27	Helmholz D	+326	-916	46	ml
XX-9	Henry, Prosper	-785	-400	42	h2,m2
XXIII-39	Heraclitus	+071	-757	90	p5,rl
XXIII-40	Heracritus D	+058	-770	52	hl,Ml,ml
XIV-19	Hercules	+434	+728	67	pl;white patch
XIV-20	Hercules A	+432	+779	33	pl;white patch
I-10	Herschel	-036	-099	41	Ml;deep floor
VII-12	Hesiodus	-245	-491	42	cl
VII-13	Hesiodus, U. F. to the N. of	-250	-472	27	ml
VII-14	Hesiodus A	-253	-501	15	cl,ml,rl;ccc
XIX-6	Hevelius	-923	+038	118	hl
I-11	Hind	+127	-138	29	M2
I-12	Hipparchus	+084	-096	151	cl,e2,ml5,p3,r6;parallel structure,ghost crater
I-13	Hipparchus, U. F. on the W. Rampart of	+048	-094	6	el
I-14	Hipparchus, U. F. on the N.-W. Rampart of	+057	-065	16	ml
I-15	Hipparchus, U. F. to the N. of	+084	-045	8	el
I-16	Hipparchus, U. F.on the S. Part of	+097	-120	10	m2
XXIV-28	Hommel	+315	-815	125	Ml
XXIV-29	Hommel A	+334	-806	51	pl
XXIV-30	Hommel C	+285	-817	53	h2

I-17	Horrocks	+102	-069	31	m1
VI-4	Hortensius	-466	+113	15	m1
XXIII-41	Huggins	-019	-656	63	h1,m1
X-11	Humboldt,W.	+878	-457	201	h3,r1,pt1
X-12	Humboldt, W., U. F. on the S.-W. of	+845	-506	47	e1
I-18	Hyginus S	+138	+112	29	m2
II-18	Hypatia C	+355	-016	16	m1

I

XXIV-31	Idler L	+262	-757	36	p1
XXI-3	Inghirami	-630	-737	91	d4,e1,r3;tcp, undulatory floor
XXI-4	Inghirami Q	-641	-742	39	M1
II-19	Isidorus B	+543	-078	30	M1,m6,p2;tcp

J

XXIII-42	Jacobi A	+144	-852	28	e1
XXIII-43	Jacobi AB	+150	-852	47	p1
XXIII-44	Jacobi C	+092	-864	35	m1
XXIII-45	Jacobi D	+090	-873	21	h1
XXIII-46	Jacobi F	+087	-853	42	m1
III-2	Jansen, U. F. on the E. of	+494	+234	42	r2;ps
II-20	Jansen B	+441	+185	16	e3;cpc
XXIV-32	Janssen	+469	-706	189	M1,r2;rf
XXIV-33	Janssen, U. F. on the N. of	+460	-683	158	M1,r1;rf
XXIV-34	Janssen E	+423	-752	25	e1
XXIV-35	Janssen J	+433	-687	30	e1
XXIV-36	Janssen X	+402	-680	24	m1,p1
II-21	Julius Caeser, U. F. to the S. of	+274	+119	24	p1

K

II-22	Kant	+340	-184	32	h2
II-23	Kant C	+370	-163	18	P1,p1
IX-35	Kant Q, U. F. on the S.-E. of	+316	-228	9	r1;ps
XI-11	Kapteyn	+927	-162	49	p2
XI-12	Kapteyn E	+924	-153	31	p1
XIX-7	Kepler	-609	+141	32	m3;rc
XXIII-47	Kinau	+127	-873	42	M1,m2

XXIII-48	Kinau D	+156	-871	27	ml
XXII-20	Kircher A	-272	-914	29	Ml
XXII-21	Kircher D	-292	-924	39	ml
VIII-42	Klein	+044	-207	44	Ml;ps
I-19	Klein, U. F. on the W. of	+027	-199	34	M4
VII-15	König	-380	-410	22	h2,rl
VI-5	Kunowsky	-536	+056	18	ml

L

VIII-43	La Caille E	+044	-398	27	pl
VIII-44	La Caille GA	+028	-344	24	dl,Ml
XXI-5	Lacroix L, U. F. on the N. of	-699	-665	35/20	ml
XXII-22	Lagalla	-273	-702	85	dl,h2,m2,Pl,pl
XX-10	Lagrange	-795	-545	160	d3,e2,M7,h5
XX-11	Lagrange E, U. F. on the E. of	-822	-486	57	h4,M2,ml
XX-12	Lagrange R	-829	-522	139	h2,Ml
I-20	Lalande	-149	-078	24	m2;mow,rc
XX-13	Lamarck A	-857	-423	56	e2,hl,Ml
XX-14	Lamarck D	-871	-425	109	dl,h6,ml,pl
V-5	Lambert	-322	+435	30	cl,hl;ps
X-13	Lamé	+873	-254	84	Ml
X-14	Lamé G	+877	-266	26	hl
XI-13	Langrenus	+863	-155	132	d2,e2,m3,p2,pt3
XI-14	Langrenus A	+906	-184	43	p2
XI-15	Langrenus K	+841	-101	29	dl,ml
VI-6	Lansberg	-448	-006	40	M2
XI-16	La Pérouse	+956	-185	78	hl,p2
XVII-3	Lavoisier A	-764	+600	27	ml
XVII-4	Lavoisier D	-737	+656	62	pl,r2
XVII-5	Lavoisier E	-745	+653	48	hl,pl
XIX-8	Letronne	-662	-184	119	Ml,ptl
VIII-45	Lexell	-059	-584	63	Ml,rl
VIII-46	Lexell A	-018	-600	33	ml,pl
XXIII-49	Lexell B	-046	-605	23	pl
XXIII-50	Licetus	+079	-733	75	h2,Ml,ml
XXIII-51	Licetus, U. F. on the S. of	+084	-744	24	pl
XXIII-52	Licetus E	+024	-704	25/18	h2
XVIII-5	Lichtenberg	-785	+527	21	e2
XVIII-6	Lichtenberg, U. F. on the N. of	-780	+537	32	ml;ps

XXIII-53	Lilius	+063 -814	61	pl
XXIII-54	Lilius A, U. F. on the S. of	+081 -829	15	ml
XXIII-55	Lilius B	+040 -798	29	hl
XXIII-56	Lilius C	+034 -813	39	ml
IX-36	Lindenau	+356 -534	53	hl,Ml,pl,r2
IX-37	Lindenau H	+379 -520	11	el
XXIV-37	Lockyer	+413 -722	34	ml
XXIV-38	Lockyer F	+402 -737	20	hl
XXIV-39	Lockyer G	+385 -715	24	ml
XXIV-40	Lockyer G, U. F. to the S. of	+385 -724	15	el
XIX-9	Lohrmann	-923 -008	34	el
X-15	Lohse	+843 -237	42	pl
XXII-23	Longomontanus	-240 -761	145	h5,M2,m2,r4;ps
XXII-24	Longomontanus A	-246 -798	29	Ml
XXII-25	Longomontanus B	-213 -797	48	ml,hl,
XXII-26	Longomontanus G	-209 -751	15	pl
XXII-27	Longomontanus Z	-206 -766	95	hl,pl
XI-17	Lubbock J, U. F. to the S. of	+649 -046	28/21	dl,m2
XI-18	Lubbock M	+625 -006	19	el
XI-19	Lubbock P, U. F. to the W. of	+622 -046	28	Ml

M

XI-20	Maclaurin	+927 -033	50	hl
XI-21	Maclaurin A	+923 -053	29	pl
XI-22	Maclaurin L	+949 -024	30	el
XI-23	Maclaurin L, U. F. to the N. of	+950 -001	22	el
XI-24	Maclaurin M	+930 -084	42	pl
XI-25	Maclaurin O	+925 -005	53	ptl
XI-26	Maclaurin P, U. F. on the E. of	+927 -100	45	rl
XII-14	Macrobius	+671 +363	64	e3,hl,pl
II-24	Mädler	+487 -191	28	pl;ps
X-16	Magelhaens A	+690 -220	30	ptl
XXIII-57	Maginus	-069 -766	162	h3,Ml,r2;innumerable hills
XXIII-58	Maginus C	-101 -785	46	h3
XXIII-59	Maginus G	-090 -743	23	ml
XXIII-60	Maginus H	-106 -793	15	el
XXIII-61	Maginus H, U. F. to the W. of	-120 -791	7	ml

XXIII-62	Maginus N	-104	-749	24	ml
XXIII-63	Maginus U	-096	-736	9	ml
XXIII-64	Maginus Z	-040	-768	14	ml,pl,ptl
XVI-4	Mairan	-514	+664	41	hl,p2;tcp
XXIV-41	Mallet B	+541	-727	33	ml
XXIV-42	Mallet D	+582	-720	42	m2
XXIV-43	Mallet J	+547	-751	52	hl,ml,rl
IV-4	Manilius	+153	+250	39	p3,rl;ps,rc
XVIII-7	Marius	-758	+206	41	ptl
XVI-5	Markov	-530	+802	41	Ml
II-25	Maskelyne	+500	+038	24	pl
II-26	Maskelyne A	+560	-001	29	Ml
XXIII-65	Maurolycus	+180	-666	114	hl,Ml
XXIII-66	Maurolycus A	+178	-688	16	ml
XXIII-67	Maurolycus D	+177	-630	45	h2
XV-15	Mayer, C.	+134	+893	38	ml,rl
XV-16	Mayer, C. B	+132	+867	36	ml
V-6	Mayer, T.	-469	+268	33	hl,r2
V-7	Mayer, T. A	-457	+263	16	hl
V-8	Mayer, T. C	-428	+212	16	cl;ccc
X-17	McClure	+742	-264	24	ml
X-18	McClure C, U. F. to the N.-W. of	+727	-249	17	sl
XXII-28	Mee	-415	-692	132	hl,M2,pl
III-3	Menelaus	+264	+280	27	h3;rc
III-4	Menelaus S	+246	+271	14	hl
XIII-2	Mercurius	+629	+726	68	ptl
XIII-3	Mercurius A	+642	+743	20	m3,pl
XX-15	Mersenius D	-670	-392	34	el,hl
XX-16	Mersenius P	-696	-341	42	hl,m2,pl
XIII-3	Messala	+670	+632	124	e2,h10,m6,p2,rl
XII-15	Messala A	+648	+596	26	hl
XIII-4	Messala D	+704	+650	28	hl
XXIV-44	Metius	+523	-647	88	ml,pl,rl
XXIV-45	Metius, U. F. to the N. of	+499	-621	23	el
XXIII-68	Miller	+010	-634	61	Ml,m2
XXIII-69	Miller A	+025	-610	39	h3;ccc
XXIII-70	Miller B	+014	-610	12	ml
XXIII-71	Miller C	-002	-619	40	hl
XIV-21	Mitchell	+223	+763	30	ml,p2,h2;ps
X-19	Monge	+697	-329	37	h2
XXIII-72	Montanari C, U. F. on the E. of	-191	-740	9	pl
XXIII-73	Moretus	-032	-943	114	h4,m3,pl

XXIII-74	Moretus, U. F. on the S. of	-020	-957	27	m2
I-21	Mösting	-101	-012	26	h1,m1;mow
XV-17	Mouchez	-092	+979	82	d1,h1,p2;rf
XV-18	Mouchez A	-080	+987	51	h2,m1
I-22	Müller	+036	-133	24/20	e1
I-23	Murchison	-002	+089	58	m6,p1,r3;plateau, heap
XXIII-75	Mutus H	+182	-896	20	h1,m1
XXIV-46	Mutus V	+237	-890	24	m1
XXIV-47	Mutus W	+255	-918	21	m1

N

XXIII-76	Nasireddin	+003	-657	51	d1,e1,h6,M1,m3,p3
IX-38	Neander	+549	-520	52	e1,m2,p2
IX-39	Neander C	+516	-479	20	e1;cpc
XXIV-48	Nearch A	+321	-867	43	m2
XXIV-49	Nearch B	+284	-874	43	p1
XI-27	Neper	+983	+155	142	M2
III-5	Newcomb	+600	+498	39	d1,h3,m2
XII-16	Newcomb J	+612	+481	23	h1
XII-17	Newcomb P	+643	+491	87/65	h2,m3,p5
XXIV-51	Nicolai Z, U. F. to the W. of	+268	-659	9	p1
XXIV-52	Nicolai Z, U. F. to the N.-W. of	+267	-645	11	m1
VIII-47	Nonius K	+058	-554	18	r2;ps
VIII-48	Nonius L	+052	-552	31	p1,r1;tcp

O

XIX-10	Olbers	-962	+124	71	e1,h1,r2
XIX-11	Olbers, U. F. to the S. of (Hedin)	-972	+054	143	e2,h3,M2,r2,p1
XIX-12	Olbers, U. F. to the S.-W. of (Hedin A)	-974	+095	63	M1
XIX-13	Olbers D, U. F. on the W. of	-968	+172	34	h1
XXIII-77	Orontius	-051	-647	109	h2

P

I-24	Palisa, U. F. to the E. of	-099	-173	40	p4;tcp,ps
X-20	Palitzsch B	+834	-445	39	p1

I-25	Pallas	-029	-096	50	m2,pl
I-26	Pallas F, U. F. to the S. of	-019	+047	10	pl
XX-17	Palmieri A	-633	-534	19	ml
VIII-49	Parrot	+057	-252	68	dl,hl,ml,P2,p3,r2;tcp
VIII-50	Parrot C	+020	-318	31	e2,Ml
VIII-51	Parrot D	+061	-245	21	Ml,pl;tcp
VIII-52	Parrot E	+038	-275	20	Ml
VIII-53	Parrot F, U. F. Containing	+017	-278	41	pl
VIII-54	Parrot J	+030	-292	23	Ml
XVI-6	Pascal	-251	+963	102	hl
XII-18	Peirce	+761	+313	18	hl
XII-19	Peirce C, U. F. on the N.-W. of	+719	+329	15	ml
XXV-7	Peirescius	+637	-725	62	el
XXIII-78	Pentland	+086	-903	56	el,pl
XXIII-79	Pentland C	+119	-906	37	hl,ml
XXIII-80	Pentland D	+110	-892	34	el
X-21	Petavius	+786	-427	177	e2,h8,M4,p2
X-22	Petavius B	+790	-340	33	el
XV-19	Philolaus	-165	+951	71	dl,hl,Ml,p3
XV-20	Philolaus D, U. F. on the S.-E. of	-104	+954	49	M5;innumerable hills
XX-18	Piazzi	-748	-590	101	dl,e3,h4;ps
IX-40	Piccolomini	+464	-497	89	e2,h3,Ml,m9
IX-41	Piccolomini C	+459	-465	26	e2,m2
IX-42	Piccolomini C, U. F. Superposed by	+453	-460	29	hl,M2,ml
IX-43	Piccolomini M	+467	-468	25	r3
IX-44	Piccolomini S	+479	-523	22	ml
I-27	Pickering	+122	-050	16	el
I-28	Pickering, U. F. on the S.-W. of	+197	-066	50	el
XXIII-81	Pictet	-093	-689	62	dl,h2,ml,pl,rl;ps
XXIII-82	Pictet A	-097	-707	34	m3;tcp
XXIII-83	Pictet A, U. F. on the S. of	-094	-718	23	pl
VII-16	Pitatus	-203	-497	60	hl,Ml,ml
VIII-55	Pitatus G	-171	-497	16	hl
VII-17	Pitatus H	-232	-508	16	hl
XXIV-53	Pitiscus	+324	-772	82	el,Ml,ml,rl
XIV-22	Plana	+350	+672	44	pl
XV-21	Plato H, U. F. to the E.of	+005	+826	7	cl;ccc

VIII-56	Playfair, U. F. to the N.-E. of	+150	-380	18	p1,m2;tcp
VIII-57	Playfair G, U. F. on the N. Wall of	+104	-385	14	pt1
III-6	Plinius	+387	+265	43	M2
XII-20	Plutarch	+897	+408	68	p1
VIII-58	Poisson, U. F. Containing	+149	-509	84	p3;tcp
VIII-59	Poisson, U. F. to the E. of	+181	-513	46	h1
VIII-60	Poisson J	+120	-574	29	h1
VIII-61	Poisson U	+152	-524	25	p3
IX-45	Polybius A, U. F. to the S. of	+434	-404	8	M1
IX-46	Pons	+332	-428	41/31	e1,h2,r1
IX-47	Pons C	+338	-469	18	m1
IX-48	Pontanus	+220	-426	54	d1,h1,M1,m2,p2;parallel linearments
IX-49	Pontanus C	+231	-502	25	p1
IX-50	Pontanus CB	+228	-497	26	p1
IX-51	Pontanus D	+206	-437	20	e1,m1
IX-52	Pontanus E	+208	-426	13	c1;ccc
IX-53	Pontanus G	+227	-510	19	m1
IX-54	Pontanus H	+237	-521	30	p1
XXIV-53	Pontécoulant	+475	-854	91	m2
XXIV-54	Pontécoulant E	+445	-870	44	e1
XXIV-55	Pontécoulant J	+428	-880	39	m1
XXIII-84	Porter (Clavius B)	-098	-830	52	M1,p1;ps
III-7	Posidonius	+424	+528	101	h4;ccc
XII-21	Proclus	+702	+278	28	h1,m3;rc
XXIII-85	Proctor,H, U. F. to the E. of	-017	-720	11	m1
XV-21	Protagoras	+071	+828	22	m1;ccc
I-29	Ptolemaeus, U. F. on the N.-E. of	+012	-138	13/9	e1
VII-18	Puiseux	-557	-466	25	h1
VIII-62	Purbach	-030	-430	118	e2,p1,r7
VIII-63	Purbach, U. F. on the N.-W. Wall of	-064	-410	8	m1
VIII-64	Purbach DB	-023	-376	19	r1
VIII-65	Purbach FA	+000	-411	14	e1
XVI-7	Pythagoras	-397	+894	128	e2,d1,h4,M2,m1,s3
XVI-8	Pythagoras, U. F. to the W. of	-430	+888	26	m1
V-9	Pytheas	-329	+351	20	h3

R

IX-55	Rabbi Levi, U. F. to the W. of	+304	-570	46	pl
VIII-66	Regiomontanus	-016	-477	126/110	h3,ml
VIII-67	Regiomontanus, U. F. to the S. of	-011	-503	12	hl
X-23	Reichenbach	+642	-504	71	ml,pl
X-24	Reichenbach A	+664	-474	34	hl
X-25	Reichenbach B	+654	-476	44	hl
X-26	Reichenbach C	+606	-490	28	ml
X-27	Reichenbach D	+621	-471	31	el,h2
X-28	Reichenbach T	+624	-490	64	p2
X-29	Reichenbach Z	+610	-529	15	ml
XIX-14	Reiner	-812	+120	30	el,Ml,pl
XIX-15	Reiner G	-814	+051	15	dl,hl,m2
VI-7	Reinhold	-387	+057	47	h4,pl;tcp
XVII-6	Repsold	-612	+779	107	Ml
XVII-7	Repsold G	-626	+771	44	Ml
XVII-8	Repsold T	-663	+739	12	Ml
XVII-9	Repsold T, U. F. to the S.-E. of	-666	+733	19	hl
I-30	Rhäticus	+086	+000	49/43	el,m2,p4
XXIV-56	Rheita	+585	-603	70	pl
IX-56	Rheita, U. F. to the N.-W. of	+579	-590	14	el
IX-57	Rheita C, U. F. to the S.-E. of	+577	-581	35	ml
X-30	Rheita E	+625	-562	66	m2
XIX-16	Riccioli	-961	-054	152	el,h6,M2,pl,r3
XXIV-57	Riccius E	+342	-642	22	el;cpc
I-31	Ritchey	+145	-193	24	pl;ps
VIII-68	Ritchey, U. F. to the S. of	+167	-205	45/25	el,h3,m4,pl,r5;cpc
II-27	Ritter	+329	+035	31	h3;ps
II-28	Ritter C	+323	+048	14	el;cpc
XX-19	Rocca A	-914	-240	60	Ml
XX-20	Rocca E	-916	-206	46	el,pl,rl
XX-21	Rocca M	-914	-252	35	h2
XX-22	Rocca Z	-929	-281	39	h2
III-8	Römer	+537	+429	40	Ml
XVIII-8	Röntgen	-839	+545	97	p2
XXIV-58	Rosenberger	+388	-823	96	el,ml,pl
XXIV-59	Rosenberger C	+412	-789	47	pl
XXIV-60	Rosenberger D	+366	-843	50	pl
III-9	Ross	+363	+202	26	el,hl

IX-58	Rothmann	+401	-513	43	r1
IX-59	Rothmann B	+405	-527	18	e1,p1
IX-60	Rothmann C	+372	-479	20	e1
XXIII-86	Rutherfurd	-102	-874	54/48	p7

S

II-29	Sabine	+343	+024	30	h3;ccc,ps
IX-61	Sacrobosco	+263	-401	96	e2,h1,M1,m3,p3;tcp
IX-62	Sacrobosco F	+270	-361	18	d1
X-31	Santbech	+650	-356	64	e1,h1,m1,p1
X-32	Santbech A	+615	-411	25	m2
I-32	Saunder	+152	-075	45	h2,m6,p2
I-33	Saunder, U. F. to the E. of	+197	-066	50	M2
XXIII-87	Saussure	-048	-686	56	e1,m1,r1
XXIII-88	Saussure CA	-006	-710	16	m1
XXII-29	Scheiner	-230	-870	110	h6,m2,r1
XXII-30	Scheiner B	-278	-862	29	m1
XVIII-9	Schiaparelli	-784	+396	24	e1
XXII-31	Schickard	-582	-700	227	d1,p2
XXII-32	Schiller	-397	-786	178	M2,m2,p1
XXII-33	Schiller, U. F. to the S. of	-390	-823	120	h3
XIX-17	Schlüter	-988	-105	91	p1
XIX-18	Schlüter A	-978	-160	35	e1
XXIII-89	Schomberger	+097	-973	85	p3
XXIII-90	Schomberger, U. F. to the N. of	+095	-964	37	d1
XXIII-91	Schomberger C, U. F. on the W. of	+042	-975	21	m2
XXIII-92	Schomberger D	+118	-959	24	e1
XI-28	Schubert	+987	+047	54	e2
XIV-23	Schwabe F, U. F. on the N. of	+317	+920	48	p1
XXIII-93	Scott	+100	-990	108	M1
XV-22	Scoresby	+052	+978	56	h1,p2;tcp
XI-29	Secchi	+688	+042	24	e1,m3
XI-30	Secchi, U. F. on the S.-W. of	+684	+033	16	e1
XXII-34	Segner	-386	-856	67	e1
XVIII-10	Seleucus	-856	+360	43	pt1
XII-22	Seneca	+878	+454	63	h1
XVI-9	Sharp	-451	+716	40	h2;ps
XVI-10	Sharp B	-485	+731	22	r3
XV-23	Sheepshanks	+150	+859	25	e1

XIII-6	Shuckburgh A	+602	+683	19	Ml;ps
XXIII-94	Short	-034	-964	71	p2
II-30	Silberschlag P	+206	+120	39/26	ml,rl
II-31	Silberschlag S	+208	+139	34/10	Ml,p5;tcp
XXIII-95	Simpelius	+077	-956	70	Ml
XXIII-96	Simpelius A	+097	-940	60	pl
XXIII-97	Simpelius D	+047	-949	54	ml
XX-23	Sirsalis	-848	-216	44	ml
XIX-19	Sirsalis E	-825	-142	72	h2
X-33	Snellius	+720	-490	83	hl
X-34	Snellius A	+716	-460	37	el
X-35	Snellius A, U. F. on the N. of	+718	-445	21	m4
X-36	Snellius D, U. F. on the E. of	+692	-479	25	hl,ml
XXIV-61	Steinheil X	+484	-738	17	el
X-37	Stevinus	+684	-537	74	Ml,m4
X-38	Stevinus B	+680	-517	20	el
X-39	Stevinus D	+637	-571	22	m2
IX-63	Stiborius	+439	-565	43	Ml
IX-64	Stiborius A	+466	-600	32	m2
IX-65	Stiborius C	+456	-558	22	el
IX-66	Stiborius EA, U. F. to the E. of	+479	-573	15	Ml
IX-67	Stiborius K	+462	-581	16	rl;cpc
IX-68	Stiborius K, U. F. to the N. of	+466	-574	10	ml
XXIII-98	Stöfler E	+073	-692	16	rl
XXIII-99	Stöfler G	+025	-687	21	el
XXIII-100	Stöfler H	+022	-648	24	Ml
XXIII-101	Stöfler J	+034	-670	73	Ml
XXIII-102	Stöfler L	+107	-631	17	hl;ps
XXIII-103	Stöfler P	+094	-687	35/27	pl,r3
XXIII-104	Street H	-140	-747	29	ml

T

IX-69	Tacitus	+313	-279	40	dl,e3,hl,m2
IX-70	Tacitus D	+348	-234	21	h2,Ml
XXIII-105	Tannerus, U. F. to the S. of	+192	-840	8	pl
XXIV-62	Tannerus C	+220	-822	16	pl
XXIII-106	Tannerus E	+188	-830	25	hl
XXIV-63	Tannerus H	+226	-811	20	el
XXIV-64	Tannerus P	+211	-825	18	ml

XI-31	Taruntius	+722	+098	56	cl;ccc,rc
II-32	Taylor	+286	-094	41/34	h2,pl
II-33	Taylor, U. F. on the N.-W. of	+301	-085	23	pl
XV-24	Theatetus	+084	+602	25	el,hl,ml;tcp
VIII-69	Thebit	-064	-375	55	el,m3,r4,pl
VIII-70	Thebit A	-079	-368	20	el,ml
VIII-71	Thebit L	-087	-366	10	ml
II-34	Theophilus	+435	-198	100	h6,M2,m7,p2
II-35	Theophilus E	+404	-118	19	h2
II-36	Theophilus F	+434	-139	12	ml,pl
XV-25	Timaeus	-004	+890	33	el,pl
V-10	Timocharis	-202	+449	35	cl
II-37	Torricelli M	+517	-064	14	el
XII-23	Tralles	+700	+476	43	el,m4
I-34	Triesnecker	+063	+073	26	m5;mow,rhombic floor,rc
XXIII-107	Tycho	-141	-685	85	el,h2,Ml,m3,p3;rc
XXIII-108	Tycho A	-159	-640	31/28	pl
XXIII-109	Tycho F	-171	-655	16	ml
XXIII-110	Tycho U	-179	-657	17	rl

U

I-35	Ukert	+024	+134	24	pl,M2;bowl-formed floor

V

XIX-20	Vasco da Gama R	-978	+172	56	rl
XXV-8	Vega B	+619	-722	30	pl
XX-24	Vieta	-726	-488	50	hl,ml
VII-19	Vitello	-525	-506	45	pl
XXIV-65	Vlacq	+374	-802	89	hl, Ml
XXIV-66	Vlacq G	+355	-818	27	ml
VIII-72	Vogel	+099	-260	27	hl
VIII-73	Vogel B	+096	-248	22	p2,rl;tcp
XVI-11	Volta	-585	+809	109	ml

W

VIII-74	Walter	+010	-545	140/132	e2,h6,M2,ml,p2,r2
VIII-75	Walter, U. F. on the Floor of	+024	-542	14	el
XXII-35	Wargentin	-565	-760	84	d4;elevated floor
XXII-36	Wargentin F	-569	-783	20	rl;cpc
XXIV-67	Watt	+487	-761	66	h3,rl

XXIV-68	Watt M	+519	-800	42	ml
XI-32	Webb	+865	-016	23	el
IX-71	Weineck	+535	-462	32	ml
VIII-76	Werner	+051	-470	70	p3,r2;tcp
VIII-77	Werner, U. F. to the E. of	+084	-462	31	ml
VIII-78	Werner H	+024	-449	16	el
XXII-37	Wilhelm A	-267	-703	20	dl;cpc
XXII-38	Wilhelm D	-227	-667	32	dl,m3
XXII-39	Wilhelm GA, U. F. on the N. of	-315	-664	27/24	rl
XXII-40	Wilhelm K	-265	-696	21	Pl;ps
XXII-41	Wilhelm V	-241	-693	8	ml
X-40	Wrottesley	+765	-405	57	Ml
VII-20	Wurzelbauer	-228	-558	82	ml

X

XVI-12	Xenophanes	-532	+842	111	hl,p4

Y

XII-24	Yerkes	+759	+252	36	ml
XXIV-69	Young D	+570	-688	46	pl

Z

XXIII-111	Zach	+045	-874	71	Ml
XXIII-112	Zach D	-065	-883	29	el
XXIII-113	Zach F	+028	-866	18	rl
IX-72	Zagut B	+272	-532	32	el,pl,rl
IX-73	Zagut C	+272	-512	24	hl,pl
IX-74	Zagut P	+253	-536	15	ml
IX-75	Zagut P, U. F. Containing	+256	-530	41	pl
XIII-7	Zeno	+674	+709	65	hl,pl
XIII-8	Zeno A	+670	+700	44	hl
XIII-9	Zeno B	+680	+695	37	Ml
II-38	Zöllner	+321	-140	47/36	h3,pl
II-39	Zöllner, U. F. to the W. of	+299	-150	19	hl;irregular formed crater
XXII-42	Zuchius	-370	-879	64	ml,pl,rl;tcp